This book describes the various techniques currently used to deposit highly ordered organic films, and the physical techniques employed in characterising their properties.

Thin organic films are the subject of a great deal of research, motivated by their potential application in micro-electronics. Existing applications, such as in the photo-lithographic production of silicon micro-circuits, involve relatively disordered films. However, higher degrees of order and the use of more complex molecules could lead to systems in which the organic molecules themselves become the active electronic devices.

Beginning with a discussion of the necessary basic physics and chemistry, the book proceeds to a description of the main topics of current research in this field. The Langmuir–Blodgett technique, self-assembly, and methods of film deposition exploiting the ordered structure of mesophases are discussed. Separate chapters are devoted to the properties and computer modelling of both liquid crystals and films at the air/water interface. Order in biomemebranes is also discussed.

The book is aimed at those graduate students and established research workers with an interest in the physics and chemistry of these fascinating structures.

ORDER IN THIN ORGANIC FILMS

ORDER IN THIN ORGANIC FILMS

R.H. TREDGOLD
Honorary Senior Research Fellow, Department of Chemistry, University of Manchester
and
Emeritus Professor, School of Physics and Materials, University of Lancaster

CAMBRIDGE UNIVERSITY PRESS
Cambridge, New York, Melbourne, Madrid, Cape Town, Singapore, São Paulo

Cambridge University Press
The Edinburgh Building, Cambridge CB2 2RU, UK

Published in the United States of America by Cambridge University Press, New York

www.cambridge.org
Information on this title: www.cambridge.org/9780521394840

First published 1994
This digitally printed first paperback version 2005

A catalogue record for this publication is available from the British Library

ISBN-13 978-0-521-39484-0 hardback
ISBN-10 0-521-39484-8 hardback

ISBN-13 978-0-521-01824-1 paperback
ISBN-10 0-521-01824-2 paperback

Contents

Contents

Preface

In writing any scientific work it is difficult to decide what background knowledge one ought to assume in potential readers. This difficulty is particularly acute when, as in this case, the book is of an interdisciplinary nature and deals with topics which belong properly to physics, chemistry and, to a certain extent, biology. When in doubt I have decided to assume ignorance rather than knowledge. For example, the text is sprinkled with diagrams illustrating the structures of chemical compounds as it is likely that readers with a physics background will be unable to deduce these structures from the names of the compounds. On the other hand, the derivation of formulae, the origins of which are readily available in common text books, has usually been omitted. When derivations are not so easily come by, they have been given. This is true, for example, in the case of the basic expression for the refractive index of a material as experienced by neutrons. I have been unable to find a derivation of this expression in any of the various books on neutron diffraction which I have examined.

The study of thin organic films has expanded enormously in recent years and it has been necessary to be very selective in order to prevent this work degenerating into a bibliography. I have attempted to discuss material in which structure and order are dealt with and to ignore the many papers, interesting from other points of view, in which these matters have not been mentioned or are given only a minor place.

The origins of diagrams and graphical results are acknowledged in detail in the figure captions but I would nevertheless like to thank once again those editors and authors who have kindly given permission to reproduce this material. Chemical formulae are the common property of the scientific community and I have not sought permission to reproduce these.

I wish to thank Professor P. Hodge for reading the entire manuscript and also to thank Dr I.R. Peterson, Dr M.J. Cook and Dr M.R. Wilson for reading parts relevant to their particular expertise. All the above have made helpful and constructive suggestions.

Professor H. Möhwald and Dr H. Matsuda both kindly provided original photographic material for which I am most grateful.

Notation

Standard international units are used throughout this book. Hydrogen atoms are not shown explicitly in chemical diagrams except where ambiguity might occur. It has not been possible to reserve a particular symbol for a particular physical quantity throughout the book as this would have meant departing from accepted usage.

1
Introductory material

1.1 The scope of this book

The major scientific advances which have taken place since the Second World War have, in some cases, had little influence on everyday life while, in other cases, they have had the most profound effect on it. Thus, for example, particle physics and astronomy have revolutionised our basic concepts of the structure of matter but have had only minor effects on the life of the non-specialist. On the other hand there are two particular fields of study whose influence has affected the inhabitants of all civilised societies. One of those is solid state electronics and the computer revolution which it has given rise to. The other is the advance in biochemistry and organic chemistry which has provided the physician with a large range of effective drugs and has transformed medicine from an art to an applied science. It is thus not surprising that the idea has arisen that a synthesis of these two fields might bring about remarkable new advances. The name molecular electronics has been proposed for this concept though different people have rather different ideas as to what is meant by this expression.

To give effect to this concept it is necessary to design and construct molecules which have certain desired physical properties and then to learn how they can be assembled in particular well ordered ways. The first problem is the province of the chemist and here considerable progress has already been made. The problem of assembly is now being studied in many laboratories using a variety of different techniques. Progress so far has been rather limited but this is hardly surprising. The only comparable problems which have been solved to date have involved the use of existing biomolecules which it took Nature a time of the order 10^9 years to produce and which have been induced to interact in the laboratory in ways similar to those in which they interact *in vivo*. Thus, even with the benefit of conscious direction, it is likely that several decades will elapse before true man-made molecules which do not simply imitate biomolecules can be made to interact reliably, even in the relatively simple ways envisaged here. It is the purpose of this book to review the techniques so far proposed to bring about such interaction and the progress made in this field.

In attempting to construct ordered arrays of organic molecules it is obvious that, at least in the first place, one should concentrate on two-dimensional or quasi-two-dimensional structures. This book thus deals with monolayers, at one extreme, and with layers a few micrometres thick at the other. For the sake of completeness, some systems are discussed where important applications exist even though the order obtained is not high. Indeed, it is precisely because the attainment of truly ordered structures is so difficult that the only practical applications of thin organic layers made so far have been those in which only a modest degree of order is necessary. Here I have in mind the spun layers of polymers used in the high resolution photo-lithography needed to produce micro-circuits on silicon crystals and also the various display devices which make use of relatively thin layers of liquid crystals.

1.2 Historical background

From a historical point of view, the progenitors of the systems we discuss are layers of paint or varnish used to preserve wood and employed for decorative purposes. The basis for these materials is normally a natural oil containing a large proportion of poly-unsaturated molecules which cross link on exposure to oxygen. A good example is tung oil employed by ancient Chinese shipwrights and discussed for example by Worcester [1]. Such layers have sufficient order to produce a continous film but are still highly random in structure. However, the study of deliberately ordered organic films is an activity which originated in the closing years of the nineteenth century. Indeed, the concept of order in a layer of material presupposes the recognition of the existence of molecules. It is thus of interest that the study of organic molecules at the air/water interface led to one of the first direct demonstrations of the molecular nature of matter by Pockels [2], as will be discussed below.

The Langmuir–Blodgett technique for deposition of ordered multilayers has origins which, in a sense, go back to classical antiquity as the present author has discussed elsewhere [3]. In 1774 Benjamin Franklin [4] demonstrated that a very small quantity of oil could influence the surface behaviour of a large area of water and thus that the basic constituent particles of the oil must be exceedingly small. However, the first truly quantitative studies of amphiphilic monolayers were made by Agnes Pockels at the end of the last century working in her own home and using simple home-made apparatus. She sent an account of her initial results to Lord Rayleigh who forwarded them to *Nature* where they were duly published [2]. She followed up this first letter by three further letters extending her results [5–7], and showed that, in the case of amphiphilic molecules such as stearic acid, there exists at the air/water interface a unique form of layer having a definite ratio of mass of stearic acid to surface area of water on which it resides. (This assertion requires some qualification and will be discussed further in Chapter 3.) In 1899 Lord Rayleigh [8] suggested that these films were monolayers and thus gave a direct measure of molecular dimensions. Using Pockels' results and the known density of stearic acid, one can deduce that these layers must be about 2.3 nm thick. Now stearic acid has a hydrophilic carboxylic group at one end of a straight hydrocarbon chain. Thus, if one assumes that the hydrophilic group lies at the water surface and that the axis of the molecule is vertical, one arrives at a value of 2.3 nm for the length of this molecule. This result compares well with the modern value of 2.5 nm for this quantity. However, what was more important at the time was the fact that the stable layer had a unique thickness and thus provided a direct confirmation of the molecular nature of matter, a concept generally accepted by chemists but still viewed with suspicion by some physicists even at the end of the nineteenth century. Subsequent work carried out by Hardy [9] and by Devaux [10] showed that only amphiphilic molecules formed good monolayers whereas simple aliphatic materials do not.

In 1917 Langmuir [11] published a systematic study of amphiphilic compounds at the air/water interface. His name has thus been attached to such films and they will be referred to as Langmuir films in this book. Nevertheless, it would be more reasonable if they had been named after Agnes Pockels or even Benjamin Franklin. In 1920 Langmuir [12] mentioned the transfer of films from the air/water interface to a solid substrate in a paper mainly devoted to other topics. In 1935 his colleague, Katherine Blodgett, published an extensive paper [13] discussing monolayers and multilayers of fatty acids deposited on a solid substrate from films existing at the air/water interface. In this book such films will be called Langmuir–Blodgett films or LB films for short. In the following 30 years a number of papers appeared dealing with the properties of such films, some of which will be discussed in Chapters 3 and 4. In 1966 Gaines [14] published a book mainly devoted to Langmuir films but also making some mention of LB films and reviewing work in that general field up to that date. At about this time Kuhn published a series of papers describing experiments in which suitably modified dye molecules were incorporated in LB films of fatty acids and their interactions studied. It was probably this work more than anything else that brought renewed interest to this field and was responsible for the growth of work on LB films which has continued throughout the 1970s and 1980s. Much of this work has been reviewed by Kuhn *et al.* [15].

The so-called self-assembly technique has its origin in a paper published in 1946 by Bigelow *et al.* [16]. These authors noted that a hydrophilic surface exposed to an amphiphilic compound dissolved in a non-polar solvent induces the amphiphilic material to form a monolayer on it. Netzer and Sagiv [17] have extended this idea to form multilayers by the following technique. A material is synthesised which has a hydrophilic group at one end and a further hydrophilic group, masked by a hydrophobic blocking group, at the other. The material is deposited on a hydrophilic surface by the method of Bigelow *et al.* [16] and the blocking groups are then removed by a chemical reaction revealing a further hydrophilic surface. Finally the whole process is repeated. In principle this ingenious process should be capable of indefinite repetition and should provide a very simple way to produce ordered multilayers. In practice there are many difficulties which will be discussed in due course in Chapter 6.

The technique of producing ordered layers of amphiphilic materials by evaporation *in vacuo* on to a suitable substrate appears to originate in the work of Agarwal [18] but has only very recently been developed.

Other methods of film formation discussed in this book depend on allowing a melt or a solution of the material to be deposited to spread on the substrate and subsequently to solidify. An ordered structure can sometimes be imposed on such a film by the application of an electric or magnetic field if the film is in a mesophase (otherwise known as a liquid crystal) before solidification. However, any such method presupposes that the melt or solution used *will* spread evenly over the substrate. It is thus important to understand a little about the conditions which allow a liquid to spread on a solid surface. This topic depends on the nature of intermolecular forces, a subject which is of general relevance to the formation of organic films and which is discussed in the following section.

The existence of liquid crystals was first observed by Reinitzer [19] in 1888, but they were first classified and examined in a systematic way by Friedel [20] in 1922. Liquid crystals and their application to the formation of ordered thin films are discussed in Chapter 7.

1.3 Intermolecular forces

The basic physics of intermolecular forces is treated by Isrealachvili [21] and the reader is referred to this book for more detailed discussion of some aspects of this topic.

The treatment which follows is of a largely qualitative nature as it seems likely that readers will fall into two classes. On the one hand there will be those who have already, at some point in their careers, familiarised themselves with the basic mathematical arguments which lead to the results quoted here. On the other hand are those who do not wish to be encumbered with lengthy algebraic manipulations but who wish to understand the basic physical mechanisms which are responsible for these results.

True covalent forces are responsible for the bonding of atoms within molecules and are usually sufficiently stable so that energies in excess of 5 eV are required to disrupt them. This energy corresponds to photons having a wavelength of less than 250 nm. In this book we are concerned with the interactions between molecules and will thus not be concerned with covalence.

The most universal and important type of intermolecular forces are known variously as van der Waals forces, dispersion forces or London forces, the latter name arising from the fact that F. London [22] gave the first proper quantum mechanical derivation of an expression for the energy arising from these forces for the case of two simple atoms. As the expression van der Waals forces is sometimes used to include other effects, we will use the term London forces here. Nevertheless, given the existence of stable molecules, London forces are not truly quantum mechanical in origin and may be understood in terms of purely classical arguments. For simplicity, consider first the interaction of two atoms which are sufficiently remote from one another so that covalent effects can be ignored. The electronic charge density of an atom will have a mean position which fluctuates with respect to the position of the nucleus. Seen from outside, this will result in a fluctuating electric dipole, the field resulting from which will tend to produce a fluctuating polarisation in the other atom, whose field will react back on the first atom. The system thus leads to a correlation between the dipolar fluctuations of the two atoms, so that there is a tendency for them to be in phase along the line joining the two atoms and in antiphase in the two orthogonal directions normal to this line. The mean dipole moments are thus always in directions which lead to an attractive force between the two atoms. For the simple case envisaged, this leads to the following expression for the London interaction energy between the two atoms.

$$W(r) = -\frac{3\alpha_1\alpha_2 I_1 I_2}{2(4\pi\epsilon_0)^2 r^6 (I_1 + I_2)} \tag{1.1}$$

Here α_1 and α_2 are the polarisabilities of the two atoms measured in C^2 m^2 J^{-1}, I_1 and I_2 are the ionisation potentials of the two atoms in joules, r is the distance between the two nuclei and the other symbols have their usual meanings. This expression is approximately correct but a number of simplifications are used in its derivation. They are as follows.

(a) In the initial Hamiltonian for the system, the expression for the potential energy associated with the interaction between the two atoms is expanded in inverse powers of r. The lowest order non-vanishing term involves the inverse cube of r, which in turn leads to the term in the inverse sixth power when a calculation of the ground state energy of the system is made. Higher order multipole terms involving all higher even inverse powers also occur in the mean energy but are usually ignored as they decrease very rapidly with increasing r.

(b) The perturbation calculation which leads to Equation (1.1) involves all the excited states of the atom having odd parity. The difference between the energies of these states and the energy of the unperturbed atomic ground state is taken to be equal to the ionisation energy. This is clearly not true but is often a reasonable approximation.

(c) In a multi-atom system, terms involving the distance apart and mutual orientations of sets of three, four and, indeed, n atoms occur but are usually ignored though, for realistic spacings in a condensed phase, these many-body terms can contribute an energy as much as 15% of the energy predicted by Equation (1.1).

It will be seen from these arguments that, even in this very simple case, the predictions of Equation (1.1) can only be accepted as approximate. In the case of real complex molecules, where the polarisability is not isotropic and little is known about the excited states, the London interaction energies as predicted by any practicable calculation must be viewed as even more approximate. It is important to bear this in mind, as many writers treat Equation (1.1) and simple elaborations of it as if it were an exact result. To obtain a feeling for the size of the effect predicted by Equation (1.1) we choose $\alpha_2 = \alpha_1$, $\alpha_1/4\pi\epsilon_0 = 1.5\times10^{-30}$ m^3 and $I_1 = I_2 = 2\times10^{-18}$ J, which are of roughly the correct size for two small molecules. Setting $r = 0.3$ nm, we obtain $W(r) = -4.6\times10^{-21}$ J, which is approximately kT at room temperature. Thus, if there are no other intermolecular forces present, materials consisting of small molecules will exist in a gaseous phase at room temperature, as indeed we find. However, as the binding energy varies as the product of the polarisabilities of the two molecules, larger molecules have interaction energies substantially larger than kT at room temperature and form liquids or solids at this temperature. Indeed, for many organic materials, London forces constitute the most important contribution to the cohesive energy.

In the standard derivation of Equation (1.1) it is assumed that the electric field experienced at one molecule arising from the charge fluctuations of the other is in antiphase with these fluctuations. This is a good approximation when the two molecules are reasonably near one another but, at larger distances, the finite velocity of light has to be taken into account. To make a rough estimate of how large such distances are, we note that the photons in question correspond approximately to the ionisation energy of the molecules and thus to wavelengths of order 200 nm or slightly less. Clearly the retardation effects will be large for molecular separations of order a quarter wavelength, so they will still be significant for distances substantially smaller than 50 nm. It may be argued that, taking into account the inverse sixth power behaviour of these forces, their behaviour at such distances is unimportant. This is true for the interaction of individual molecules but, when one considers the interaction of condensed phases, London forces at such distances can still be of importance. The retardation effect was first treated properly in a quantitative manner by Casimer and Polder [23]. It is clear that the effect will be to make the interaction fall off more rapidly than the inverse sixth power of r at larger distances.

Many of the other important attractive forces between molecules can be explained in terms of electrostatics. The simplest such force leads to an attractive energy of interaction between spherically symmetrical monovalent ions of unlike sign given by the familiar expression

$$W(r) = \frac{-q^2}{4\pi\epsilon_0\epsilon r} \tag{1.2}$$

where ϵ is the relative permittivity of the medium in which the ions are immersed and q is the electronic charge.

The energy of interaction of a monovalent ion and a spherically symmetrical uncharged molecule having a polarisability α is given by

$$W(r)=\frac{-\alpha q^2}{2(4\pi\epsilon_0\epsilon)^2r^4} \tag{1.3}$$

where again a medium of relative permittivity ϵ is assumed. Clearly a real molecule cannot be described as a sphere, but the error in treating one as such is in many cases small.

The energy of interaction of a fixed dipole and a polarisable spherically symmetrical molecule depends on the angle θ between the dipole axis and the line joining the centre of the molecule and the dipole. For a dipole of moment u, the field at a distance r and at an angle θ is given by

$$E=\frac{u(1+3\cos^2\theta)^{\frac{1}{2}}}{4\pi\epsilon_0 r^3} \tag{1.4}$$

and the interaction energy is thus given by

$$W(r,\theta)=\frac{-u^2\alpha(1+3\cos^2\theta)}{2(4\pi\epsilon_0\epsilon)^2r^6} \tag{1.5}$$

It will be seen that this energy varies as the sixth power of r, as does the energy arising from London forces. Here the reason for this variation is rather more obvious. The field arising from the dipole varies as the inverse cube of the separation and thus so will the induced dipole on the molecule. This molecular dipole will in turn produce a field which varies as the inverse cube of r at the original dipole.

For two fixed dipoles the energy of interaction is given by

$$W(r,\theta_1,\theta_2,\phi)=\frac{u_1u_2(\sin\theta_1\sin\theta_2\cos\phi-2\cos\theta_1\cos\theta_2)}{4\pi\epsilon_0\epsilon r^3} \tag{1.6}$$

where θ_1 and θ_2 are the angles between the line joining the dipoles and the axis of dipole 1 and dipole 2 respectively which have moments u_1 and u_2. Here again the medium is supposed to have a permittivity ϵ. ϕ is the azimuthal angle between the dipoles.

Certain of these formal results can be used in a straightforward way to make important qualitative predictions about the behaviour of molecular systems. Consider, for example, a mixture of two liquids 1 and 2, whose molecules can be approximated to spheres and which rely largely on London forces for their cohesive energy. Clearly Equation (1.1) can be written approximately as

$$W_{12}(r) = \frac{-\beta_1\beta_2}{r^6} \tag{1.7}$$

If the liquids are treated as disordered close-packed solids each molecule has, on average, twelve neighbours. Suppose further that there are SN molecules of type 1 and $(1-S)N$ molecules of type 2. From Equation (1.7) it follows that

$$W_{12}^2(r) = W_{11}(r)\,W_{22}(r) \tag{1.8}$$

If we take only nearest neighbour interactions into account (which is not too bad as a first approximation in the case of an interaction which falls off as the inverse sixth power of r) and if we assume that the mixture is perfectly random, then the internal energy (neglecting surface effects) is

$$U_{\text{random}} = \frac{1}{2}SN[12SW_{11} + 12(1-S)W_{12}] + \frac{1}{2}(1-S)N[12(1-S)W_{22} + 12SW_{12}] \tag{1.9}$$

The arrangement of the molecules which minimises the London energy leads to an expression for this energy summed over the whole system, which is given approximately by

$$U_{\text{ordered}} = 6N[SW_{11} + (1-S)W_{22}] \tag{1.10}$$

The difference between these two expressions is

$$U_{\text{random}} - U_{\text{ordered}} = \frac{6NS}{r^6}(1-S)(\beta_1 - \beta_2)^2 \tag{1.11}$$

and thus the internal energy is always lowest for the case when the two liquids are in separate phases. As the temperature is raised, the entropy term becomes progressively more important and eventually the free energy becomes a minimum for a random mixture. This is a valid but simplified account of the behaviour predicted by more sophisticated treatments and corresponds to the behaviour of many real systems.

Electrostatic forces contribute a major part to the hydrophilic and hydrophobic effects which are important in the discussion of Langmuir–Blodgett films and other topics dealt with in this book. Water is the commonest and also probably the most complex terrestrial liquid. It is clearly a simplification to treat it as an assembly of individual molecules each bearing an electric dipole moment. Nevertheless, such a description is true enough for our present purposes. If a molecule which incorporates an ionic charge or an electric dipole is introduced into water, the water molecules rearrange themselves in the vicinity of this molecule so as to lower the electrostatic energy of the system. A molecule which does not contain either an unpaired charge or an electric dipole and which is introduced into water forces apart the water molecules which were formerly arranged in such way as to lower their electrostatic interactions. Thus this situation increases the internal energy of the system and it is unfavourable from an energetic point of view to introduce this molecule into water. This simple argument would predict that liquids should be either infinitely miscible or immiscible with water depending on whether they are polar or not. However, the entropy term in the free energy has been neglected and when it is included the possibility of partial miscibility arises. Furthermore, it is usually claimed that the entropy term contributes to the hydrophobic effect. The argument is that water molecules arrange themselves in an ordered way round a hydrophobic molecule and thus decrease the entropy of the system and hence increase the free energy.

It remains to discuss hydrogen bonding which can contribute a bonding energy of between about 5 and 20 kT between suitable pairs of molecules at room temperature. Examples are the bonds which hold water and also ice together and those which attract the oxygen atoms to the NH groups in adjacent protein chains. This latter effect is also important in attracting oxygen atoms to NH groups further along the same protein chain and thus, when other conditions are favourable, stabilising the α-helix structure. In all these cases a hydrogen atom attached by a covalent bond to an electronegative atom such as O, N, F or Cl is attracted to a further electronegative atom. Unlike simple covalency where the true quantum mechanical phenomenon can be explained in terms of a convincing hand-waving picture, it is difficult to give a valid qualitative picture of the hydrogen bond. The reason for this is that, to give a quantum mechanical explanation of hydrogen bonding, it is necessary to use a wave function which corresponds to a mixture of different configurations of the system.

So far we have only discussed attractive effects. The origin of the repulsive force between molecules having a closed shell structure is easy to visualise but complicated to describe in a quantitative manner. When two such molecules are brought into contact the electrons on one start to invade the region corresponding to the orbitals of the electrons on the other molecule. This is a contravention of the exclusion principle. If however, electrons are excited to higher states, their orbitals can still occupy the same region of space. Thus, to bring the two molecules closer together, work has to be done to promote electrons to excited states. Naturally this verbal description is over simple and really one must consider mixtures of configurations, in some of which electrons exist in excited states. However, the basic mechanism is easy to visualise and it is evident why repulsive forces only become important when molecules approach one another closely.

1.4 Wetting of solid surfaces

The relation of surface tension to wetting was first discussed by Young [24] in 1805 and was also treated by Laplace at about the same time. Modern developments in this field have been reviewed by Zisman [25] and more recently by de Gennes [26]. Surface tension has the dimensions of force per length and is measured in dynes cm^{-1} or $mN\ m^{-1}$ the numerical values being identical for the two systems of units. Surface energies, which can be visualised in terms of the integration of a constant surface tension over unit distance, have the same numerical value as the corresponding surface tension. In this book we use the symbol γ to denote surface tension with, where appropriate, a double subscript to denote the two phases between which the surface in question lies. Reference to Figure 1.1 shows that, for a liquid drop on a solid surface, the equilibrium situation is given by

$$\gamma_{SV} - \gamma_{SL} - \gamma_{LV} \cos \theta = 0 \qquad (1.12)$$

The subscripts S, L and V correspond to solid, liquid and vapour respectively. This is a classical result which is valid on a macroscopic scale. On a microscopic scale various other factors arise in the region of the junction of the three phases. This microscopic behaviour is treated by de Gennes [26] but will not be examined further here.

When

$$\gamma_{SV} - \gamma_{SL} - \gamma_{LV} = 0 \qquad (1.13)$$

$\cos \theta = 1$ and total wetting occurs. As γ_{SL} increases beyond the point which satisfies Equation (1.13) complete wetting still persists. One can

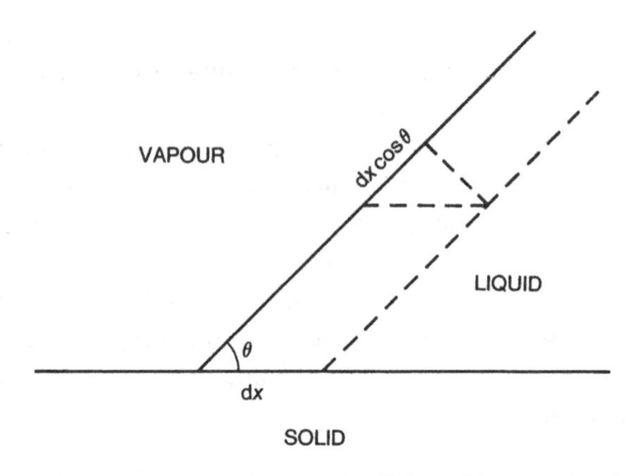

Figure 1.1. Derivation of Equation (1.12). When this equation is satisfied an infinitesimal change of position of the three-phase point, dx, will not change the total surface energy of the system and thus it is in equilibrium.

see in a qualitative way the kind of situation which will lead to complete wetting. For simplicity consider a non-polar liquid in which the molecular interaction within the liquid and between the liquid and the solid is dominated by London forces. If the high frequency refractive index and hence the high frequency permittivity of the solid is greater than that of the liquid, liquid/solid attractions will be greater than liquid/liquid attractions and the lowest energy state will correspond to the liquid spreading out to cover as much of the solid surface as possible. When hydrogen bonding is present the situation will be complicated and the relative strength of hydrogen bonds within the liquid and between the liquid and the solid must also be taken into account. Thus surfaces which have an ionic character are likely to be hydrophilic and surfaces of materials where covalent bonds or London forces predominate are likely to be hydrophobic.

Returning to the case in which London forces predominate we note that, in the case that the high frequency permittivity of the liquid is larger than that of the solid, the liquid will try to form droplets, the contact angle will be greater than zero and partial wetting will take place. For a series of otherwise similar liquids the contact angle will thus be a function of the permittivity of the liquid, a parameter which is in itself related to the surface tension of the liquid. If contact angle is plotted as a function of surface tension and the plot is extrapolated to the point where the contact angle is zero, a value of surface tension is obtained which is often the same for different series of liquids. Zisman [25] was the first individual to note this fact. A linear plot of surface tension of the liquid versus $\cos\theta$ is thus known as a Zisman plot. As de Gennes [26] points out, there is no reason to postulate a linear relationship between these two quantities and other workers have employed other functional relationships. The value of surface tension corresponding to zero contact angle in such a plot is supposed to characterise the surface in a unique way but it is clear that this will only be so when the various molecular interactions all predominantly arise from London forces.

There is a hysteresis effect associated with the wetting process which arises, at least in part, from the presence of defects and impurities in the surface under study. Thus advancing and retracting contact angles are slightly different. It is conventional to use the advancing contact angle in the Zisman plot.

2
Definitions of order and methods of its measurement

2.1 Definitions of order

In the last chapter we have used the word 'order' without giving it any precise meaning. Most definitions of order involve thermodynamic concepts. Thus, for example, one might say that the most ordered state of a system is the one to which the system tends as the temperature tends to absolute zero. This definition would, however, be of little service in the present context. Most of the systems which we will discuss are remote from thermodynamic equilibrium. This is true both of the films during their preparation and also of the 'final' prepared films. However, these prepared films are in states of metastable equilibrium which are likely to survive for periods long compared with the time taken to carry out experiments on them and, very often, for periods so long as to be, from a human point of view, infinite.

We thus need a different definition of order. Here it is suggested that we view the most ordered state as the one which corresponds most closely to some preconceived structure which we wish to bring about. From a practical point of view the extent to which we can tolerate disorder may vary widely depending on the context. Thus, for example, in a system analogous to a chain of DNA encoding the structure of a particular enzyme, a single defect may render the system useless. On the other hand, a nematic liquid crystal display unit may perform its function if the axis of each molecule in a particular region lies at an angle not too far removed from that of the director and no other order exists.

In conventional solid state physics it is usual to view a perfect single crystal as the ideal ordered system and a state related to this perfect crystal but containing some dislocations and localised lattice defects as the most highly ordered state attainable in practice. In the materials that we are to discuss, such a degree of perfection is rarely achieved. The nearest approach to true three-dimensional long range order that we are likely to arrive at would be obtained by epitaxial growth on the surface of a macroscopic three-dimensional crystal. At the time of writing, very limited progress in this direction has been made using organic molecules. This is hardly surprising as it is difficult to obtain as a substrate three-dimensional crystalline material which can be cleaved to produce a step-free surface, which is not destroyed by the process of deposition and which has a suitable lattice constant to accommodate organic materials. Most of the thin films of organic materials achieved to date consist either of polycrystalline structures in which true three-dimensional crystalline structure exists only over very small microscopic areas or are really best described as frozen liquid crystals. Indeed, some Langmuir–Blodgett films are true smectic liquid crystals even at room temperature.

As a broad generalisation, the systems which we are here concerned with can be thought of as ordered in three different ways.

(a) Multilayers repeat their structure in a more or less regular way in a direction normal to the plane of the layers and this regularity can be investigated by X-ray diffraction and also by neutron diffraction.

(b) Within a particular layer it is sometimes possible to obtain a two-dimensional crystalline structure, but a hexatic structure is more usually found and this form of structure is discussed in Section 3.4.

(c) The degree to which rod-like molecules all share the same axial direction is also of major interest. The mean direction of these axes, known as the director vector, is discussed further in Chapter 7. The extent to which the axes of individual molecules depart from this direction is often measured by an order parameter, S, which is defined as

$$S=\frac{1}{2}\int_0^\pi \rho(\theta)\,(3\cos^2\theta-1)\sin\theta\,\mathrm{d}\theta$$

where $\rho(\theta)$ is the probability of the axis of a particular molecule making an angle θ to the director at an arbitrary time. It is difficult to investigate S in the case of a thin multilayer but, in the case of hydrated bilayers of phospholipids where a large number of bilayers alternate with thin layers of water, nuclear magnetic resonance may be employed and this topic is discussed in Chapter 8.

2.2 Methods of measuring the degree of order

Further discussion of the concept of order is best related to the description of the various experimental techniques used to measure this quantity in thin organic films. The list given below is not exhaustive but includes most of the more important techniques which have been employed.

1. Probably the most important of these is the diffraction of waves from the film. X-ray diffraction, neutron diffraction and electron diffraction have all been employed and examples of their use will be discussed below. For films less than a few micrometres thick, X-rays and neutrons can only be used in the low angle mode as the diffracting power of such films is not sufficient to obtain useful results if the incident beam is directed normally at the film. Electron diffraction can be used with relatively thin films but the destructive effect of the high energy beam often damages the film before an image can be recorded.

2. Optical and electron microscopy are again basic and important techniques in this context but, once again, the latter can cause rapid film deterioration.

3. The scanning tunnelling microscope (STM) is such a newly available tool that it is difficult to assess its true value in the study of organic films. Under favourable conditions it can provide atomic resolution and, with a tip to substrate potential of less than 1 V, it is unlikely to do damage. It is limited in its application to films one or two monolayers thick and, at the present state of development, it is not always clear what one is really seeing or why one is seeing it. Atomic force microscopy (AFM) uses techniques in some ways analogous to STM. A fine tip is scanned over the material to be examined, against which it is held by a force of between 10 and 100 nN. Its movements are magnified by an optical lever and recorded. It is thus possible to build up an image of the surface of a material which conducts electrons so poorly as to render the use of STM ineffective. AFM is an even more recent development than STM but it seems likely to be of more use than the former technique when applied to organic films, many of which are very poor conductors of electricity. Very recent results appear to confirm this statement.

4. Infrared, visible and near ultraviolet spectroscopy are particularly important when used in conjunction with polarising devices in determining the mean orientation of molecules or particular parts of molecules.

5. Electrical measurements of various kinds give information about the macroscopic properties of films which can be of assistance in the study of structure. These include measurements of capacity versus number of layers, measurements of conductivity both across the plane

of the film and through it and studies of apparent surface potential versus thickness.

6. For the upper end of the thickness range which we will consider, mechanical measurement of thickness using one of the commercially available stylus devices can provide valuable information.

7. Ellipsometry is also an important technique for thickness measurement.

2.3 Diffraction from a nearly perfect lattice

The theory of the diffraction of waves from a crystal has been extensively discussed in a variety of texts and the basic formulae will not be re-derived here. Reference will be made where appropriate in this book to the classic work by James [27]. The formulae given here are, however, presented using SI units rather than CGS units as in the original work. The problem of X-ray diffraction from liquid crystals, in many ways analogous to problems to be discussed here, has been treated at some length by Leadbetter [28]. In the succeeding paragraphs a few of the more important relationships concerning this topic will be exhibited and relevant results derived.

Let 2θ be the angle through which rays are scattered and λ be the wavelength. Define $|\mathbf{Q}| = 4\pi \sin\theta/\lambda$ where the direction of the vector, $\mathbf{Q}$, is the difference between the wave vectors corresponding to the scattered wave and the incident wave. If A is the amplitude of the scattered wave and f_p is the scattering factor for the pth unit cell whose origin is situated at the point $\mathbf{r}_p$ then the intensity of the scattered wave is given by

$$A^2 = CI \tag{2.1}$$

where

$$C = \frac{1}{R^2}\left(\frac{e^2}{4\pi\epsilon_o Mc^2}\right)^2 \frac{1+\cos^2 2\theta}{2} \tag{2.2}$$

and

$$I(Q) = \sum_{p,q} f_p f_q^* \exp[\mathrm{i}\mathbf{Q}\cdot(\mathbf{r}_p - \mathbf{r}_q)] \tag{2.3}$$

Here R is the distance of the detector from the specimen, the trigonometric factor in Equation (2.2) accounts for the effects of polarisation and the other factors have their usual physical meanings. The amplitude of the incident wave is taken to be unity, f is a function of Q which depends on the electron distribution within the unit cell. These results are derived in most standard texts on X-ray diffraction and the derivation will not be repeated here.

Clearly it is Equation (2.3) which contains all the interesting information about the lattice and the scattering process. Using this expression one can develop many of the standard formulae relating to the diffraction of X-rays from a crystal and closely analogous expressions apply to neutron and electron diffraction.

Here we wish to examine the simple one-dimensional case and discuss the effect of order on diffraction. The one-dimensional equivalent of Equation (2.3) is

$$I(Q)=\sum_{p,q} f_p f_q^* \exp[\mathrm{i}Q(x_p-x_q)] \tag{2.4}$$

It is convenient and legitimate to replace the sums by integrals and write

$$I(Q)=|f|^2 \int_{x_p}\int_{x_q} \rho(x_p)\rho(x_q)\exp[\mathrm{i}Q(x_p-x_q)]\,\mathrm{d}x_p\,\mathrm{d}x_q \tag{2.5}$$

where $\rho(x)$ is a density function which is normalised so that the integral of this function taken over the length of the one-dimensional 'crystal', L, equals the number of atoms in crystal. f is now a constant related to the number of electrons in each unit cell which take part in the scattering process. Thus

$$\int \rho(x)\,\mathrm{d}x=N \tag{2.6}$$

If the atoms are supposed to be located at points then

$$\rho(x)=\sum_{n=0}^{N-1}\delta(x-nd) \tag{2.7}$$

where d is the lattice constant and $\delta(x)$ is the delta function which is zero everywhere except at $\delta(0)$ and has the property

$$\int F(x)\delta(x-x_0)\,\mathrm{d}x=F(x_0) \tag{2.8}$$

where $F(x)$ is any function and the range of integration in (2.8) includes the point $x=x_0$.

Here in Equation (2.7) n is an integer defining a particular lattice site. Using (2.5) and (2.7) it is trivial to recover (2.4). However, the advantage of using (2.5) is that we can include the description of structure within the unit cell without additional complication and we will make use of this in Section 2.4.

Returning to (2.4) and putting $x_p = n_p d$ etc, one can obtain for a finite lattice containing N unit cells

$$I(Q) = |f|^2 \left[N + 2 \sum_{n=1}^{N-1} (N-n) \cos Qnd \right]$$

$$= |f|^2 \frac{\sin^2 QdN/2}{\sin^2 Qd/2} \tag{2.9}$$

by straightforward manipulation. (See [27] page 522.) The second form of the intensity function in this equation is more generally familiar but the Fourier series leads directly to the concept of the reciprocal lattice (discussed in Section 2.5) and is also convenient as a limiting form with which to compare the effects of disorder.

If one examines the series in Equation (2.9) it will be seen that the terms will tend to reinforce whenever $Q = 2\pi l/d$, where l denotes an integer. These values of Q define a one-dimensional reciprocal lattice and whenever Q takes on one of these values the diffracted waves will reinforce. These values of Q correspond to the familiar Bragg reflections given by $2d \sin \theta = l\lambda$. It should be noted that for finite values of N the reciprocal lattice 'points' have a finite width.

So far we have used the scattering factor, f, to take account of the effect of structure within the unit cell. It is instructive to make use of Equation (2.5) instead. At one extreme the case of all the scattering power in a unit cell being located at one point leads to Equation (2.9) with f being a constant. In this extreme case, all orders of diffraction have the same amplitude. As the electron density within the unit cell is represented by more spread out and realistic functions, so the higher order terms in Equation (2.9) decrease more rapidly with increasing values of Q. Thus in the extreme case represented by

$$\rho(x) = \frac{1}{L} \left[\exp\left(\frac{2\pi i x}{d}\right) + \exp\left(\frac{-2\pi i x}{d}\right) \right] \tag{2.10}$$

and $f = 1$ we obtain from (2.5) the result

$$I(Q) = \delta\left(Q - \frac{2\pi}{d}\right) + \delta\left(Q + \frac{2\pi}{d}\right) \tag{2.11}$$

which means that diffraction only occurs in association with the lowest order reciprocal lattice points. (Here it has been assumed that the integrations in (2.5) are taken over such large distances as compared with the lattice spacing that it is legitimate to treat the 'fuzzing out' of reciprocal lattice points as unimportant.) Real cases correspond, of course, to a situation somewhere between these two extremes. It is beyond the scope of this book to give a complete discussion of the methods whereby the form of $\rho(x)$ can be calculated given the intensity of the various orders of diffraction but relevant special cases will be treated in Section 2.4. What we are concerned with here, however, is the influence of order on the intensity of these successive orders of diffraction.

To simplify the discussion we will restrict it to the case of 'point' atoms and to one dimension and thus return to Equation (2.4). Suppose that x_p has a random error u_p superimposed on it so that

$$x_p = dn_p + u_p \tag{2.12}$$

with a similar relationship for q. If it is assumed that $Qu \ll 1$ then, by expanding the exponentials and taking means, one can show that, to a good approximation,

$$I(Q) = \sum_{p,q} f_p f_q^* \exp\left[\mathrm{i}Qd(n_p - n_q)\right] \exp(-Q^2u^2) \tag{2.13}$$

where u is the mean error in the position of an atom and it is assumed that these errors do not have any correlation as between neighbours. Using a similar calculation to that which leads to Equation (2.9) we obtain

$$\begin{aligned} I(Q) &= |f|^2 N\left[1 + 2\sum_{n=1}^{N-1}\left(1 - \frac{n}{N}\right)\mathrm{e}^{-u^2Q^2}\cos(Qnd)\right] \\ &= |f|^2\left[N(1 - \mathrm{e}^{-u^2Q^2}) + \mathrm{e}^{-u^2Q^2}\frac{\sin^2(QNd/2)}{\sin^2(Qd/2)}\right] \end{aligned} \tag{2.14}$$

This calculation assumes that long range order exists in the system and that the 'errors' are simple thermal deviations from the positions defined by this order. This diminution of intensity with increasing Q was first discussed by Debye [31, 32] and Waller [33] and is hence known as the Debye–Waller effect.

In this book we wish to discuss partially disordered systems and cannot thus always expect true long range order to exist. A proper treatment of the statistical mechanics of a partially ordered system is extremely difficult, though some success has been achieved by numerical modelling and will be referred to in Chapter 7. It is thus important to understand the behaviour of the limiting cases which can be analysed. Clearly one limiting case corresponds to the structure having long range order on which a small random perturbation is superimposed, which has been discussed above. The other limiting case involves the superposition of successive relative errors as one proceeds along a chain.

Consider Equation (2.5) and suppose that the distance over which any sensible correlation of atomic positions exists is much less than the length of the chain. It is then a good approximation to write

$$I(Q)=|f|^2N\int_{-\infty}^{+\infty} s(x)\mathrm{e}^{\mathrm{i}Qx}\,\mathrm{d}x \tag{2.15}$$

where $s(x)$ is now the probability of finding an atom at a distance x from an arbitrarily chosen 'origin' atom. Indeed, this expression is just the one-dimensional form of the intensity function usually used to calculate X-ray diffraction from a simple liquid.

The probability of finding the 'origin' atom at $x=0$ is then just

$$P_0(x_0)=\delta(x_0) \tag{2.16}$$

Suppose now that the probability of finding the next atom (which resides at a mean distance d in a positive direction) at a position x is

$$P_1(x)=\phi(x-d) \tag{2.17}$$

where

$$\phi(x)=\frac{\exp[-x^2/(2\sigma^2)]}{(2\pi)^{1/2}\sigma} \tag{2.18}$$

and generally that

$$P_n(x_n)=\int P_{n-1}(x_{n-1})\phi(x_n-d-x_{n-1})\,\mathrm{d}x_{n-1} \tag{2.19}$$

In particular

$$P_1(x)=\frac{\exp[-(x-d)^2/(2\sigma^2)]}{(2\pi)^{1/2}\sigma} \tag{2.20}$$

$$P_2(x)=\frac{\exp[-(x-2d)^2/(4\sigma^2)]}{(2\pi)^{1/2}\sigma\sqrt{2}} \tag{2.21}$$

By successive convolutions we obtain

$$P_n(x)=\frac{\exp[-(x-nd)^2/(2n\sigma^2)]}{(2\pi)^{1/2}\sigma\sqrt{n}} \tag{2.22}$$

Now

$$s(x) = \sum_{n=-\infty}^{+\infty} P_n(x) \tag{2.23}$$

From (2.15) and (2.23) we obtain

$$I(Q) = |f|^2 N \sum_{n=-\infty}^{+\infty} \int P_n(x) e^{iQx}\, dx \tag{2.24}$$

Using (2.16) with (2.22) and (2.24) we obtain

$$I(Q) = |f|^2 N \left[1 + 2 \sum_{n=1}^{\infty} \exp\left(\frac{-Q^2\sigma^2 n}{2}\right) \cos(Qnd) \right] \tag{2.25}$$

It is interesting to compare this result with (2.14) which assumes long range order whereas (2.24) does not. One cannot expect 2.25 with $\sigma = 0$ to correspond to (2.14) with $u = 0$ as we have employed an approximation to obtain (2.15). For finite values of σ and u, the most obvious difference in behaviour between the functions in (2.14) and (2.25) is the fact that all orders of diffraction defined by the latter equation are widened as well as being decreased in magnitude by the exponential factor and that this widening increases rapidly as one proceeds to higher orders of diffraction.

It is worth pointing out that, even without the effects of disorder, there will be a fairly rapid, though not monotonic, decrease of intensity with progressively higher order of diffraction maxima except in the artificial case of point atoms. The extreme case of this effect is illustrated by the argument associated with Equations (2.10) and (2.11).

The simple arguments presented above are for a one-dimensional variation of order in a direction normal to the planes of a layer structure in which individual layers remain flat. Such a system is, of course, a gross over-simplification of a real three-dimensional system but nevertheless serves to illustrate the way in which deterioration of order influences the X-ray diffraction obtained from a layered structure such as is obtained, for example, by the Langmuir–Blodgett technique.

2.4 X-ray diffraction

The basic laws of X-ray diffraction are embodied in the equations in Section 2.3. If one simply wishes to predict the position of diffraction maxima and not their intensity and width, the use of the simple Bragg law is adequate.

$$2d \sin \theta = l\lambda \tag{2.26}$$

Here d is the interplanar spacing, l is an integer and λ is the wavelength. θ is the angle between the direction of the incident X-ray beam and the lattice plane under consideration and also, of course, the angle between the direction of the diffracted beam and this plane. For films up to a few micrometres thick (the region of concern to us in this book) a beam of X-rays normal to the film intercepts too thin a layer of material to give an easily detected diffracted beam. We are thus limited to aiming the incident beam in a direction only a few degrees from the tangent to the film. This low angle X-ray diffraction is limited in scope in that it can readily give us information about the spacings of layers which lie in the plane of the film and about the corresponding variation in electron density normal to these planes but is ill adapted to yield other information. Layer spacing can be obtained with considerable accuracy. A plot of $\sin \theta$ versus $l\lambda$ can very often yield d accurate to within 0.2%.

The apparatus usually employed to carry out this process involves an X-ray generator producing copper Kα radiation at a wavelength of 0.1541 nm which is collimated and aimed to impinge tangentially on the film in its initial position. The film is usually deposited on a glass slide or on some other amorphous material which will not produce coherently diffracted radiation which would interfere with the radiation diffracted from the film itself. The film is then slowly rotated through a uniformly increasing angle, θ, about a line which lies in the plane of the film and is normal to the direction of the incident radiation. A diffracted beam of X-rays will be produced whenever the Bragg condition is satisfied and this beam will travel in a direction which lies at an angle 2θ to the incident radiation. The mechanical system which rotates the specimen is coupled to a system which rotates a radiation detector through an angle 2θ. The output from this detector is fed via an amplifier to the y-input of a pen-recorder whose x displacement is proportional to θ and varies linearly with time. The $00l$ diffraction peaks are thus directly displayed as a function of θ. Some modern diffractometers use rather different means to bring about these ends and store all the information in a computer but the final result is the same. Holley and Bernstein [29, 30] were the first workers to obtain such diffraction patterns from Langmuir–Blodgett films and subsequently a large number of papers have been published in this field some of which will be discussed briefly below.

Results which arise from films having long range order in the direction normal to the plane have chiefly been obtained from Langmuir–Blodgett multilayers formed from long-chain carboxylic acids deposited from a subphase containing a divalent cation, usually cadmium, though other types of molecule and other methods of deposition have also led to ordered layer structures which can be studied by this technique. The high electron density associated with cadmium is particularly helpful in obtaining strong diffraction peaks. Given that the layers of molecules point alternatively up and down with respect to the film plane (the so-called Y structure) one would expect that the thickness of one repeat unit would be roughly equal to the length of two molecules placed end to end. Indeed, this is often *roughly* the case but there are four factors which can prevent this result being exact.

(a) In the case of carboxylic acid films formed from a subphase containing multivalent cations one must allow for the space occupied by the cations.
(b) For many materials, stable multilayers form in which the hydrocarbon chains lie at an angle (often about 30°) from the normal to the film plane.
(c) In the case of some materials having a rather more complex structure than those so far mentioned, it is possible for molecules to interdigitate and thus reduce the repeat distance.
(d) It is usually assumed that the hydrocarbon chains involved are in an all-*trans* configuration but this need not always be the case.

Simple measurements of Bragg layer spacing are not sufficient to distinguish (b) from (c) especially as the complication arising from (a) may also be present. To obtain unambiguous knowledge of the structure of a multilayer one must either make use of other data arising, for example, from electron diffraction or one must carry out a proper structure analysis using the relative intensities of the peaks corresponding to a number of orders of diffraction.

If we are only interested in the variation of electron density in the direction normal to the plane of the film, a quantity defined in Section 2.3 as f_p, we can treat f as a constant and expand $\rho(x)$ as a Fourier series.

$$\rho(x) = 1/L \sum_{n=-\infty}^{+\infty} a_n \exp\left(\frac{2\pi inx}{d}\right) \tag{2.27}$$

where L is the film thickness.

To proceed further without undue complication we need to make two approximations.

(a) It must be assumed that the film is sufficiently thick that we can ignore the effects of finite thickness on the diffraction pattern. In practice this means that one requires a film consisting of at least 50 repeat units.

(b) One must ignore the Debye–Waller correction for the influence of temperature. The effect of this neglect is to underestimate the influence of the higher Fourier components in Equation (2.27) and thus reduce the resolution of detail. A proper allowance for the Debye–Waller effect would be extremely difficult to make for the case of a film consisting of relatively large molecules and will not be discussed here.

Given these approximations and using Equations (2.5) and (2.27), one obtains

$$I(Q) = |f|^2 \sum_{n=1}^{\infty} |a_n|^2 \left[\delta\left(Q - \frac{2\pi n}{d}\right) + \delta\left(Q + \frac{2\pi n}{d}\right) \right] \qquad (2.28)$$

where the term involving $Q=0$ has been neglected. This zeroth-order term coincides with the large direct X-ray beam and so cannot be measured in any case. Given that one is dealing with a centrosymmetric structure. $a_n = a_{-n}$ and is purely real. However, these quantities can be positive or negative and this sign cannot be deduced directly from the experimental results which correspond to Equation (2.28). The problem of finding the signs of these Fourier coefficients is the celebrated phase problem and is central to the determination of crystal structures by X-ray diffraction. In the case of the quasi-one-dimensional structure we are considering and given that we know the structure of the molecules comprising the film, this problem can usually be solved by a rather laborious but straightforward process. The repeat distance, d, and the absolute values of the Fourier coefficients can be found directly from the diffraction pattern. It is thus possible to write a computer programme which calculates and displays the various different possible forms of $\rho(x)$ depending on the signs attributed to the coefficients. Suppose that there are nine Bragg peaks which are sufficiently large compared with the noise that they are worthy of consideration. The sign of the first is arbitrary so that one has $2^8 = 256$ possibilities to consider. Before the advent of powerful desk top computers the examination of all these possibilities was an impractical task and a variety of ingenious methods have been devised to circumvent it. (See for example the work of Lesslauer and Blasie [34].) However, it is not difficult to programme a computer having a graphics facility to generate and store the various forms of $\rho(x)$ and to display them sequentially. There remains the problem of comparing these functions with possible distribution functions and eliminating the large number which are implausible. Tredgold *et al.* [35] used a process whereby the monitor screen was divided into four quarters each of which displayed one of the possible forms of $\rho(x)$. Forms of this function which, on initial examination, appeared compatible with the physical situation were then retained and each of these was re-examined on the whole screen. Various simple guides can be used in this selection process. For example, in layer structures arising from most of the materials belonging to the general class discussed in this section, there are substantial regions consisting of simple saturated hydrocarbon chains lying parallel to one another. $\rho(x)$ should remain almost constant over such regions. The ultimate selection of the preferred set of phases is a process of inductive reasoning in which all possible influences such as tilt and interdigitation must be examined. In the study cited [35], layer structures formed from two similar compounds differing only in the length of a straight chain region were available, a factor which provided extra information.

Similar structural studies have been made, for example, by Feigin and Lvov [36] using Y layers of barium behenate and also superlattices containing this material and other amphiphiles. Structural studies of Langmuir–Blodgett films have also been made by Belbeoch *et al.* [37].

The Langmuir–Blodgett method has been used by various workers to study X-ray diffraction from systems consisting of only a few monolayers in which the finite thickness of the sample is of paramount importance. An interesting example of such work has been published by Pomerantz and Segmuller [38].

It is impossible in a work of this kind to list all the papers in which X-ray diffraction has been used to study thin organic films. Particular papers will be cited as they bear on the study of specific materials. A selection of papers referring to Langmuir–Blodgett films is listed [38–47].

2.5 Electron diffraction

Just as X-ray diffraction is best adapted to the study of the structure of a film in the direction normal to the plane of the film, the X-rays impinging nearly tangentially to it, so electron diffraction is applicable to the converse case where the electrons impinge in the direction of the normal and provide information about the two-dimensional structure of the film in its plane. For typical operating conditions the diffracting power of a thin film for electrons is of the order 10^4 times greater than the diffracting power for X-rays. On the other hand, random scattering and inelastic losses of various kinds are so large that transmission electron diffraction is only applicable to films having a thickness less than about 0.1 μm. The two techniques are thus complementary and it is thus unfortunate that they are rarely applied to the same materials in the same laboratory.

The study of X-ray diffraction in thin films is essentially a one-dimensional problem and so we have been able to avoid any awkward geometry in that case. This is not true for the case of electron diffraction and it is thus worthwhile introducing the reciprocal lattice explicitly. We return once more to Equation (2.3)

$$I(Q) = \sum_{p,q} f_p f_q^* \exp[\mathrm{i}\boldsymbol{Q} \cdot (\boldsymbol{r}_p - \boldsymbol{r}_q)] \tag{2.3a}$$

If it is possible to choose $\boldsymbol{Q}$ so that all the exponents are zero or $2\pi n\mathrm{i}$, where n is an integer, then the terms on the right hand side will sum so as to reinforce and represent a diffraction maximum. We thus seek values of $\boldsymbol{Q}$ which will bring this about.

Let the primitive translations in the real lattice be the vectors $\boldsymbol{a}$, $\boldsymbol{b}$, $\boldsymbol{c}$, then these quantities multiplied by the integers u, v, w, respectively define the lattice points. Thus

$$\boldsymbol{r}_p = \boldsymbol{a}\, u_p + \boldsymbol{b} v_p + \boldsymbol{c} w_p \tag{2.29}$$

with a similar relationship for $\boldsymbol{r}_q$.

In a similar manner let the primitive translations in the reciprocal lattice be the vectors $\boldsymbol{a}^*$, $\boldsymbol{b}^*$ $\boldsymbol{c}^*$ corresponding to the integers h, k, l. (The asterisk here denotes the reciprocal lattice, not a complex conjugate.)

The reciprocal lattice vectors are defined implicitly by the equations

$$\boldsymbol{a}^* \cdot \boldsymbol{b} = \boldsymbol{a}^* \cdot \boldsymbol{c} = \boldsymbol{b}^* \cdot \boldsymbol{c} = \boldsymbol{b}^* \cdot \boldsymbol{a} = \boldsymbol{c}^* \cdot \boldsymbol{a} = \boldsymbol{c}^* \cdot \boldsymbol{b} = 0 \tag{2.30}$$

and

$$\boldsymbol{a}^* \cdot \boldsymbol{a} = \boldsymbol{b}^* \cdot \boldsymbol{b} = \boldsymbol{c}^* \cdot \boldsymbol{c} = 1 \tag{2.31}$$

One then finds that if

$$\boldsymbol{Q} = 2\pi(h\boldsymbol{a}^* + k\boldsymbol{b}^* + l\boldsymbol{c}^*) \tag{2.32}$$

then

$$\boldsymbol{Q} \cdot (\boldsymbol{r}_p - \boldsymbol{r}_q) = 2\pi n \tag{2.33}$$

which is the required relationship.

Thus, if $\boldsymbol{Q}$ corresponds to the vector joining the origin to any other point in the reciprocal lattice multiplied by 2π, then a diffraction maximum will be observed. If one is dealing with a finite real lattice, the points in the reciprocal lattice will be blurred and will occupy a finite region of reciprocal space. In particular, when one is considering a thin film, the regions in reciprocal space corresponding to diffraction will be

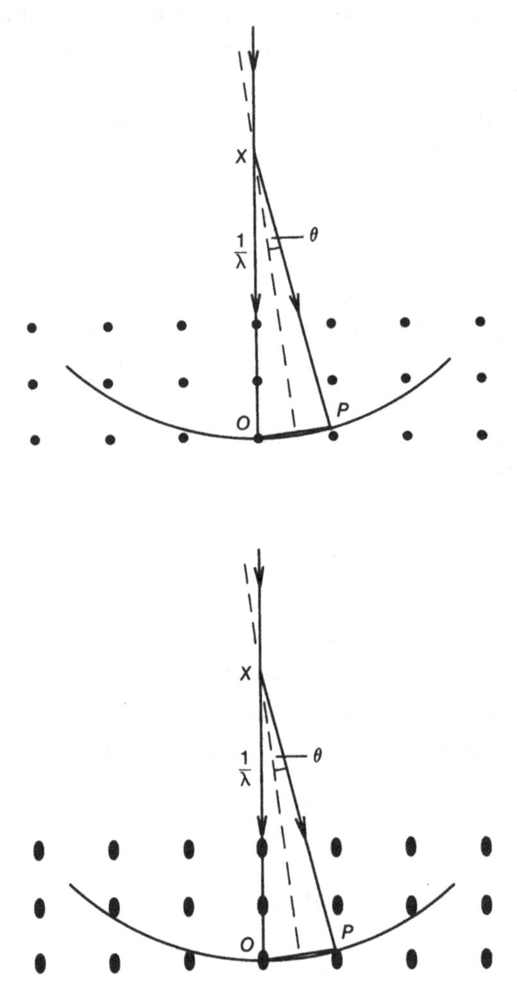

Figure 2.1. Application of the reciprocal lattice to the analysis of electron diffraction data. (*a*) The vector corresponding to the incident wave is drawn through the origin of the reciprocal lattice, O, in the direction that the wave is travelling and has a length, XO, equal to $1/\lambda$. For diffraction to take place Q must correspond to 2π times the vector joining O to another point in the reciprocal lattice, P, and distance XO must be equal to XP. Clearly this situation can only be satisfied by 'freak' conditions. However, (*b*) illustrates what happens for the real case of a thin film in which the reciprocal lattice 'points' become extended in the direction normal to the plane of the film.

extended in the direction normal to the plane of the film. Figure 2.1(*a*) illustrates this concept in a schematic manner and is a two-dimensional example of a form of diagram first introduced by Ewald [48] and von Laue [49]. The vector corresponding to the incident wave is drawn through the origin of the reciprocal lattice in the direction in which this wave is travelling and has a length proportional to $1/\lambda$. For diffraction to take place, $\boldsymbol{Q}$ must correspond to the vector joining the origin to another point in reciprocal space multiplied by 2π. Thus diffraction maxima will only occur when the incident wave vector has one of the special values which brings this about. It may be shown that these special conditions are the conditions for a Bragg reflection. Now consider Figure 2.1(*b*) which represents the reciprocal lattice produced by a thin film. It is now no longer necessary for the incident wave vector to be in a special direction in order to produce a diffracted wave, though a maximum in diffracted intensity will still be obtained when the Bragg condition is fulfilled. Thus, if one is considering a thin film, diffraction will occur at points corresponding to a projection of the reciprocal lattice on a plane normal to the incident wave vector, the pattern being scaled by a factor which involves the wavelength and the geometry of the instrument being used. Electron diffraction experiments are usually carried out using suitably focused electron microscopes with a beam energy of about 100 keV. This corresponds to a wavelength of 0.0037 nm which is of the order 30 times less than typical lattice spacings in a simple inorganic materials and, for large organic molecules, the factor is considerably bigger. Thus $1/\lambda$ is very much bigger than the reciprocal lattice spacings and the arguments presented in connection with Figure 2.1(b) are justified.

Further discussion of this subject is given, for example, by Loretto [50].

2.6 Neutron diffraction

The neutron being without charge is not scattered by Coulomb forces but may be scattered by short range forces associated with the nuclei of the atoms of the material through which it is passing or by the magnetic moments of unpaired electrons. The second of these two phenomena is of importance in the study of ferromagnetic and antiferromagnetic materials but is not of direct relevance to the subject matter of this book. Neutron diffraction has a considerable literature and a number of books have been devoted to it, one of the more recent of which was edited by Willis [51]. The subject will only be discussed here in so far as it applies to the characterisation of layer structures formed from organic molecules. The importance of neutron diffraction in this context is that the nuclei of hydrogen and deuterium scatter neutrons differently and thus the deuteration of particular molecules or parts of molecules can be used as a tool to investigate the way in which they have been deposited. In order to explain how this is done it is necessary to give a very brief discussion of the theory of the diffraction of neutrons by crystals.

Let us first examine the scattering of a neutron plane wave by a single atom, which is supposed fixed. The behaviour of this system is represented by the wave equation

$$-\frac{\hbar^2}{2m}\nabla^2\psi+V(r)\psi=E\psi \tag{2.34}$$

Here $V(\boldsymbol{r})$ is the potential energy of interaction of the neutron and the nucleus of the atom.

Let the incident wave be represented by $\exp(\mathrm{i}\boldsymbol{k}_0\cdot\boldsymbol{r})$. Using the Born approximation (the Born approximation is explained in most standard texts on quantum mechanics) the approximate solution of this equation is given by

$$\psi=\psi_0+\psi_1$$

$$\psi_1=\frac{-1}{4\pi}\int\frac{\exp[\mathrm{i}\boldsymbol{r}'\cdot(\boldsymbol{k}_0-\boldsymbol{k}_1)]\exp(\mathrm{i}\boldsymbol{r}\cdot\boldsymbol{k}_1)\,2mV(\boldsymbol{r}')}{|\boldsymbol{r}-\boldsymbol{r}'|\,\hbar^2}\,\mathrm{d}\tau' \tag{2.35}$$

where $\boldsymbol{k}_1$ is the wave vector for the scattered wave. Now the region of interaction is confined to a volume only a little larger than the dimensions of the nucleus (of the order 10^{-14} m) in diameter. Thus for large $\boldsymbol{r}$ we can approximate $|\boldsymbol{r}-\boldsymbol{r}'|$ to $\boldsymbol{r}$. Furthermore, since we are only interested in very small values of $\boldsymbol{r}$ we can approximate $\exp\mathrm{i}[\boldsymbol{r}'\cdot(\boldsymbol{k}_0-\boldsymbol{k}_1)]$ to 1. For the scattered wave we then obtain

$$\psi_1=\frac{-\exp(\mathrm{i}\boldsymbol{k}_1\cdot\boldsymbol{r})}{4\pi r}\int\frac{2mV(\boldsymbol{r}')}{\hbar^2}\,\mathrm{d}\tau' \tag{2.36}$$

For an incoming wave having unit amplitude the scattered neutron flux in a solid angle $\mathrm{d}\Omega$ is

$$\mathrm{d}\sigma=b^2\,\mathrm{d}\Omega \tag{2.37}$$

For the approximations used above, b is given by

$$b=\frac{1}{4\pi}\int\frac{2mV(\boldsymbol{r}')}{\hbar^2}\,\mathrm{d}\tau' \tag{2.38}$$

The total flux is given by

$$\sigma = 4\pi b^2 \tag{2.39}$$

When one considers scattering by more than one atom there is the additional complication that the quantity denoted here by b is different for different isotopes and that nuclear spin is also involved in the process. So far as isotope effects are concerned, we will only be concerned with systems where the different isotopes, hydrogen and deuterium, are in predictable positions and so there is no need to analyse the effects of their random distribution in detail. However, in the case of both hydrogen and deuterium, the nuclear spins of the different atoms are, under normal circumstances, randomly arranged. Now the proton has spin $\frac{1}{2}$ and so does the neutron. Thus the combined system can be in a triplet or in a singlet state and the effective value of b is different for the two cases.

Let us denote the the mean value of b for a mixture of isotopes by $\langle b \rangle$ and let $\Delta b(\mathbf{r})$ be the fluctuation from this mean for a particular value of $\mathbf{r}$. Then, using the symbol $\langle \ \rangle$ to denote averaging over all values of $\mathbf{r}$

$$b(\mathbf{r}) = \langle b \rangle + \Delta b(\mathbf{r}) \tag{2.40}$$

Thus as $\langle \Delta b(\mathbf{r}) \rangle = 0$

$$\langle (\Delta b)^2 \rangle = \langle b^2 \rangle - \langle b \rangle^2 \tag{2.41}$$

and thus the scattered neutron flux is given by

$$\begin{aligned} \sigma &= 4\pi \langle b^2 \rangle \\ &= 4\pi \langle b \rangle^2 + 4\pi(\langle b^2 \rangle - \langle b \rangle^2) \end{aligned} \tag{2.42}$$

where the first term on the right hand side represents the coherent scattering and the second term the incoherent scattering.

For the particular case of hydrogen, the relative probability of the triplet state is ¾ and of the singlet state it is ¼.

The values of b for these two cases are given by

$$\begin{aligned} b_{\mathrm{T}} &= 1.075 \times 10^{-14}\,\mathrm{m} \\ b_{\mathrm{S}} &= -4.738 \times 10^{-14}\,\mathrm{m} \end{aligned} \tag{2.43}$$

Hence, for hydrogen,

$$\begin{aligned} \sigma_{\mathrm{CO}} &= 1.8 \times 10^{-28}\,\mathrm{m}^2 \\ \sigma_{\mathrm{INCO}} &= 79.7 \times 10^{-28}\,\mathrm{m}^2 \end{aligned} \tag{2.44}$$

In the case of deuterium, the nuclear spin is 1 and the combined spins of the neutron and the nucleus can be ½ or ¾. Using an argument analogous to the one given above for hydrogen, one arrives at

$$\sigma_{CO} = 5.6 \times 10^{-28}\,\mathrm{m}^2$$
$$\sigma_{INCO} = 2.0 \times 10^{-28}\,\mathrm{m}^2 \tag{2.45}$$

The fact that the coherent scattering cross section of deuterium is three times that of hydrogen can be exploited to distinguish the regions which are deuterated from those which are not, as will be discussed below. In principle. one could develop the scattering theory for neutrons exactly as has been done for X-rays and electrons. However, the experimental studies which will be cited have used the technique of deuterating alternate layers of molecules and, in order to relate to them, it is better to think of the material under study as consisting of alternating slabs of material having different refractive indices for neutrons. As the velocity of thermal neutrons is much less than that of light, it is valid to think of a refractive index greater than unity for cases when the neutron velocity in the bulk material is less than that in free space and refractive indices *less than unity* when the velocity in the bulk material is greater than that in free space. To obtain an expression for this index of diffraction in terms of b and the number of atoms per unit volume we use an ingenious approximation developed by Fermi [52]. Figure 2.2 represents the edge view of a thin sheet of material (the thickness is exaggerated for diagrammatic purposes) on which a neutron plane wave is incident from the left. It is supposed that the sheet is sufficiently thin so that attenuation of the wave in it may be ignored, that there are N atoms m^{-3} each having a scattering length, b, and that the incident wave can be represented by the function e^{ikz}. Consider a sheet of thickness $d\omega$ located at $z=\omega$. Using (2.36)–(2.38) we can write for the incident and scattered wave

$$\psi = \exp(ikz) - bN \int_0^t \exp(ik\omega)\,d\omega \int_{\rho=0}^{\infty} \frac{\exp(ikr)}{r} 2\pi\rho\,d\rho \tag{2.46}$$

where the other symbols are defined in Figure 2.2. Using the fact that $\rho\,d\rho = r\,dr$, this becomes

$$\psi = \exp(ikz) - 2\pi bN \int_0^t \exp(ik\omega)\,d\omega \int_{r=z-\omega}^{\infty} \exp(ikr)\,dr \tag{2.47}$$

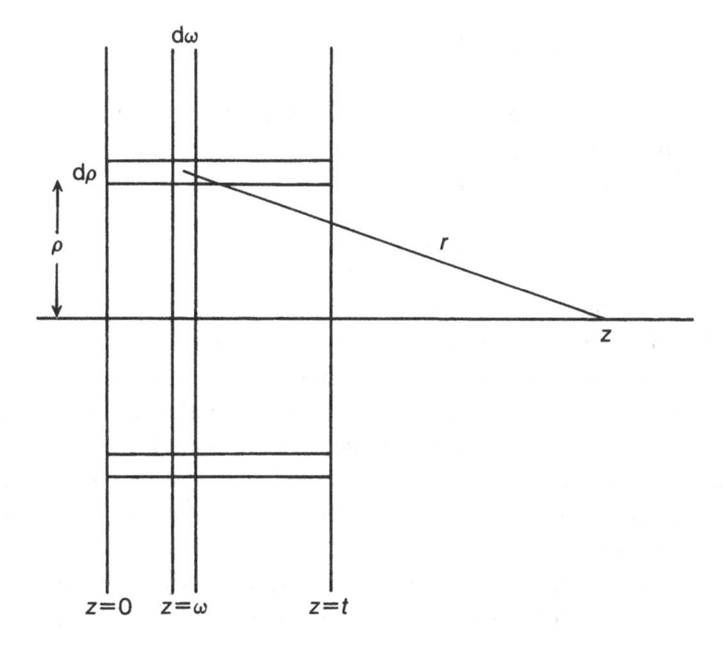

Figure 2.2. Derivation of the standard neutron diffraction expression, Equation (2.51). It is supposed that neutron plane waves proceed from the left and can be represented by the function e^{ikz} and impinge on a slab of material whose faces are defined by $z=0$ and $z=t$. Consider an infinitesimal slice of this material defined by $z=\omega$ and $z=\omega+d\omega$ and a ring of this slice confined between ρ and $\rho+d\rho$ where ρ is the radius of this ring. The wave function at z, a point on the axis of the ring and to the right of the slab will now, using the equations in the text, be given by Equation (2.48). Using the manipulations shown in the text, one then arrives at Equation (2.51).

The integral is indeterminate at the upper limit but the introduction of a convergence factor reduces the value at the upper limit to zero. As Fermi points out, this is a reasonable procedure as the beam intensity will decay at its edges. We thus obtain

$$\psi=e^{ikz}\left(1-\frac{2\pi bNti}{k}\right) \tag{2.48}$$

Now, if the refractive index is denoted by n, simple optic theory shows that the outgoing beam will be given by

$$\psi=\exp(ikz)\exp[ikt(n-1)] \tag{2.49}$$

Equating the expressions obtained from (2.48) and (2.49) and dividing by e^{ikz}, we have

$$1 - \frac{2\pi bNti}{k} = \exp[ikt(n-1)] \tag{2.50}$$

As n is only slightly larger than 1 we can expand the exponential term in (2.49) and retain only the first two terms. Simple manipulation and the use of the identity $k = 2\pi/\lambda$ leads to

$$n = 1 - \frac{\lambda^2 bN}{2\pi} \tag{2.51}$$

As we have assumed that the region over which neutron interaction with the nucleus is appreciable is much smaller than the neutron wavelength, we are dealing with a true quantum phenomenon and semiclassical methods are not valid. It is interesting that a calculation which assumes that the nuclear potentials are averaged over the bulk and in which semiclassical methods are adopted leads to a result which has a superficial resemblance to (2.51) but with a plus sign before the second term on the right hand side.

For real cases there will be more than one kind of atom present in any one region and it will be necessary to use a mean value of b, but the fact remains that, for alternate layers of deuterated and undeuterated material, different values of the refractive index will be effective in the deuterated and undeuterated regions.

Broadly speaking there are two distinct experimental methods employed in neutron diffraction.

In the first method, thermal neutrons from a nuclear fission reactor are collimated and then monochromated by diffraction from a lead crystal. After further collimation they are diffracted by the material under study and detected by a neutron counter. In an alternative version of this method monochromation is accomplished by a velocity filter consisting of slits in two discs rotating on the same axle. In either case the basic geometry is similar to that employed in X-ray diffraction. The incident ray strikes the specimen at an angle θ with respect to the crystal planes and, when the Bragg condition is satisfied, is reflected at this angle. Thus, just as in X-ray diffraction, the detector has to be positioned at an angle 2θ with respect to the incident beam and a mechanical system is employed which scans the specimen through an angle θ and the detector through an angle 2θ at the same time. The density of neutrons is low relative to the density of quanta in an X-ray beam and it is thus only possible to operate in the low angle mode employing a slow scan rate.

In the second method, high density pulses of neutrons are generated by a spallation source. In this case an initial compact pulse of neutrons containing particles having a wide variety of velocities, and hence a wide variety of wavelengths, is generated. The specimen and detector are located at fixed values of θ and 2θ respectively but the time of flight of those neutrons which are diffracted can be measured with precision. Thus the velocity and hence wavelength can be found.

To summarise, the first method involves a fixed wavelength and a variable θ while the second involves a fixed θ and a variable λ. In either case the purpose is to find the conditions for Bragg reflection. The advantage of the second method is that higher neutron densities can be obtained and faster experiments and higher levels of resolution can be achieved.

It is evident from physical considerations that the Bragg conditions for diffraction will be of exactly the same kind as for X-rays. However, quantitative results require the solution of the optical problem of diffraction by a finite stack of dielectric sheets having a periodic variation of dielectric constant. This problem is best handled by numerical means and the question is discussed by Hayter *et al.* [53], who also give references to earlier papers in which the numerical methods are fully discussed. Similar methods have also been made use of by Nicklow *et al.* [54]. Further studies were made by Highfield *et al.* [55].

The most important use of neutron diffraction in the general field encompassed by this book is in the study of alternating layers of deuterated and undeuterated films. At the time of writing, three papers have appeared on this topic. They are by Buhaenko *et al.* [56], Grundy *et al.* [57] and Stroeve *et al.* [58]. In the latter study, alternate layers of deuterated and undeuterated fatty acids were deposited and studied by neutron diffraction. Subsequently this ordered structure was destroyed by thermal diffusion and the gradual loss of order was monitored. The ordered structure is only destroyed at temperatures well above ambient. Applications of neutron diffraction to the study of lipid films at the air/water interface will be discussed in Chapter 8.

2.7 Optical measurements

X-ray diffraction, electron diffraction and neutron diffraction all involve using radiations whose wavelengths are of the same order as the distances between the molecules. However, it is possible to obtain valuable information about the state of order of a thin organic film using radiations in the visible and infrared regions. A material which has anisotropic optical properties in the plane normal to the direction of the incident light will exhibit this structure when studied using a polarising microscope. As a very rough empirical guide, it is usually necessary to have about 30 or more molecular layers before this property of the material can be made use of. The two phenomena used are birefringence and dichroism. In the case of birefringence, the polariser and analyser are set at right angles to one another and the specimen is rotated in a plane normal to the axis of the microscope. Maxima in the transmission of light appear at relative angles of the rotating stage of 0, $\pi/2$, π and $3\pi/2$. It is thus often possible to decide if a material is homeotropic or has some kind of domain or polycrystalline property. In some cases other interesting structural characteristics appear; these will be discussed later in this book.

Dichroism is the term applied to the variation of optical absorption with the direction of polarisation of the incident light. In this case, only a polariser *or* an analyser is used and transmission of light is a maximum for relative angles of the microscope stage of 0 or π. In this case, transmission depends on angle and also on wavelength. In the ideal case it would be possible to illuminate the specimen using a monochromator and study dichroism as a function both of the position of the specimen and of wavelength. In practice it is not usually possible to obtain adequate illumination this way. Where large regions of the specimen are oriented in the same direction, it is possible to abandon microscopy and use a spectrometer fitted with a polarising attachment to investigate dichroism. Clearly the study of dichroism by microscopy is only possible where there is a reasonably strong absorption band in or near to the visible region.

The study of infrared dichroism using polarised light is a powerful tool in the study of thin films of organics. A modern Fourier-transform infrared spectrometer allows one to integrate the results arising from successive scans for lengthy periods and also to subtract absorption effects arising from the substrate on which the specimen is mounted. Under favourable circumstances it is thus possible to obtain useful results from a single molecular layer. The specimen must be deposited on a material which, as far as possible, absorbs only weakly in the infrared region in which important and well characterised absorption bands occur. This region corresponds approximately to wavelengths ranging from 3 μm to 10 μm. Proprietary materials consisting of glasses made using chalcogenides are often employed. It is then possible to examine absorption in the case that the direction of propagation is normal to the plane of the specimen and the electric vector lies in the plane. If there exists anisotropy in the plane of the specimen and a polarising attachment is fitted, then this anisotropy can be investigated.

Using a complementary technique, it is also possible to study infrared absorption when the electric vector is normal to the plane of the specimen. The technique employed was originated by Allara *et al.* [59] and involves depositing the organic film on a metal layer which has been deposited *in vacuo* on a microscope slide. The radiation impinges on the slide at a glancing angle and is polarised so that the electric vector is nearly normal to the plane of the slide. The combination of the incoming and reflected waves produces a resultant electric vector normal to the surface. Thus, by combining these two techniques, it is possible to find the angles which various bonds make with respect to the plane of the film.

A third method of investigating the infrared absorption spectra of thin organic films is the attenuated total reflection (ATR) technique. In this method a piece of germanium or silicon cut in the form of a thin slab with the thin opposing edges cut at an angle of about 45° to the plane of the slab is used. These particular semiconductors have a relatively low number of absorption bands in the infrared region of interest. The radiation is launched via one of the bevelled edges and at such an angle that it suffers total internal reflection at one of the larger plane surfaces. A series of internal reflections then take place and the radiation is finally emitted at the other bevelled edge. Now, when total internal reflection takes place there exists an evanescent wave immediately outside the solid material. Thus, if the semiconductor slab is coated with a thin film of the organic material to be studied, it will be penetrated by this wave and energy will be absorbed at wavelengths corresponding to the usual absorption spectra of the material. The repeated reflections will magnify this effect. It should be noted that the semiconductive property of the slab used is not made use of and that these materials are employed simply because they are available as large single crystals having little absorption over a reasonable part of the relevant infrared region.

A number of special optical techniques such as the study of second harmonic generation by irradiation of non-centrosymmetric systems by high intensity laser light will be discussed in relation to particular materials and problems. However, one optical technique having a general applicability, namely ellipsometry, must be discussed here. It is one of the best techniques available to determine the thickness of a thin organic film. Such determinations are important as they allow one to have an independent check on the number of layers deposited, given that the thickness of one layer has been determined by X-ray diffraction.

Ellipsometry is unusual in that it depends on classical electromagnetic theory but was only developed in the 1960s. A linearly polarised light wave is reflected from a dielectric surface on which a thin layer of another insulating material has been deposited. The wave can be viewed as a superposition of two waves, of which one has the electric vector in the plane containing the wave vectors of the incident and the reflected waves and the other has the electric vector at right angles to this plane. It is traditional to denote the first of these by the letter p and the second by the letter s. The phase changes resulting from reflection differ for these two waves and this difference in phase depends on the thickness of the thin layer and its refractive index. A derivation of the expressions needed to calculate this thickness from the settings of a polariser associated with the incident wave and an analyser in the path of the outgoing wave set so as to prevent the passage of light (and including a quarter wave plate intercepting either the incident or outgoing wave) is straightforward but beyond the scope of this book. The reader is referred to the *Handbook of Optics* [60] for further discussion and citations of the basic papers on this subject.

Ellipsometry is suitable for the measurement of films whose thickness is much less than the wavelength of light. The thickness measurement of films having hundreds of molecular layers is best carried out using a stylus device such as the Talystep. Here a groove is scribed in the film and the stylus measures the depth of this groove. In the case of very soft materials, the film can be coated with a metallic layer having a thickness of about 20 nm which is deposited *in vacuo* after the line has been scribed.

The characterisation techniques discussed in this chapter are of a fairly general nature. Specialised techniques applicable to particular problems are discussed as the questions arise in the text.

3
Films at the air/water interface

3.1 Trough technology

Amphiphilic materials spread at the air/water interface have been the subject of intensive study over a long period of time. The type of apparatus usually used for this purpose has much in common with the apparatus needed to form Langmuir–Blodgett films and, indeed, it is usually possible to adapt the same apparatus for both purposes. In this section the problems which must be overcome if these processes are to be carried out are discussed and the most effective solutions to these problems described.

It has become traditional to use the word 'trough' to denote such apparatus and this usage will be adhered to here though the word trough tends to suggest such things as hogwash rather than the ultra-cleanliness needed for effective studies of monolayers. It is indeed this cleanliness which must be our first concern. Many materials which would otherwise be suitable for the fabrication of troughs tend to leach out surface active material into the water based subphase contained in the trough and thus can not be used. Most modern troughs are made in one of the two following ways.

1. Teflon (polytetrafluoroethylene) does not leach out plasticisers and can be purchased in substantial blocks, sheets and rods of various thicknesses. The preferred method is to machine a trough from a solid block of this material. This is a practicable procedure if a good milling machine is available but one is limited to a rather shallow trough. This does not matter if it is to be used for the study of monolayers at the air/water interface only but is a severe limitation if it is to be used to make LB films. Attempts have been made to use a metal trough and coat it using a Teflon aerosol. Such attempts have not proved very successful.

2. The other popular way of making troughs is to use borosilicate glass. A good scientific glass blower can make a trough having a semicylindrical form with closed ends. If a very large trough is required (say 1 m long with the other dimensions in proportion) a different procedure is called for, as it is difficult to make a hard glass trough of this size in one piece. The way to proceed is to form a semicylindrical trough without closed ends and to grind the ends flat. Separate flat end plates with Teflon gaskets can then be used and, if they are held in place with a little pressure, there is no leakage. The only difficulty which arises in this case is that cleaning procedures are made more complicated.

As a high degree of cleanliness must be maintained, the trough must either be in a clean room or in a glove box. If appropriate precautions are taken, the latter method leads to a higher level of cleanliness and is, of course, much cheaper. Ordinary glove box gloves are not really suitable as they are packed in French chalk and the resulting dust drops into the trough. Long (arm length) veterinary gloves are ideal. The glove box should have a large transparent lid which can be opened for major operations such as cleaning the trough or changing the subphase. To eliminate dust, one must flush the glove box with air which has been filtered down to sub-micrometre levels. A closed circuit system gives the best results but, if the same quantity of air is continually recirculated over the trough, the relative humidity rises rapidly and the deposition of LB layers becomes impossible. To avoid this problem, the duct which carries the air can be passed through a cold zone provided, for example, by an ordinary domestic deep freeze unit, which freezes out the moisture.

The amphiphilic material to be studied is dissolved at a known concentration in a volatile solvent which is not miscible with water and a known quantity is spread at the water surface using a micropipette. In order to study the physical properties of the film thus formed, one needs to be able to confine the film to a definite area and to be able to vary this area at will. It might appear that it would be equally possible to maintain a constant area and vary the amount of material which is spread. For the majority of materials this latter procedure is not satisfactory as equilibrium is not arrived at in a reasonable period of time and this method would not allow one to take the material through successive cycles of compression and expansion. We thus turn to a discussion of the various ways in which a film can be confined and its area varied in a systematic manner. Leaving aside methods which are really only of historical interest, for which reference should be made to the book by

Gaines [14], there are two basic methods of confining and compressing films at the air/water interface.

1. The edges of the trough are machined flat and are maintained in a horizontal position. They are either made of a hydrophobic material such as Teflon or are covered with a hydrophobic material such as wax. Only the former procedure is normally now used. The trough is filled to a point where it almost overflows. If a Teflon barrier is arranged so that it lies flat on, and makes contact with, both sides of the trough it will, under these circumstances, act as a barrier to the film which is spread between it and one end of the trough.
2. The constant perimeter trough was originally introduced by Blight *et al.* [61] and was subsequently developed by Roberts *et al.* [62]. A closed loop of Teflon coated tape passes round a series of pulleys and intercepts the water surface. By a suitable mechanical system arranged to move the pulleys, it is possible to change the area enclosed by the tape while the perimeter remains constant. Figure 3.1 gives a schematic diagram of such a system without showing the mechanical drive. It will be evident that two of the three sets of pulleys must be mounted on carriages which move in synchronism so as to vary the perimeter and maintain tension in the tape. This result is usually achieved by mounting the pulleys on carriages which are supported by small wheels running on rails and which are moved by a system of belts which, in their turn, are moved by an electric motor via a reduction gear. This system involves mechanical complication but is symmetrical. A simple system which achieves nearly the same result is shown in Figure 3.2. Here only one carriage has to move but the region of tape between the two mobile pulleys moves sideways and can cause problems.

The constant perimeter technique is in many ways the most effective way of manipulating a film at the air/water interface but leads to a laborious amount of cleaning as the tape and the pulleys (made of Teflon) must be removed and cleaned in a Soxhlet apparatus at regular intervals.

The difficulty in using either of these methods is that the monolayer tends to adhere to the sides of the trough and thus distorts under compression. This problem was indeed recognised by Blodgett [13] who suggested a solution to it, a solution which has been revived recently by Kumehara *et al.* [63]. In this method the film is confined at one end by a floating barrier which is attached to two waxed silk ribbons which intercept the water surface and form the sides of the enclosed area. These

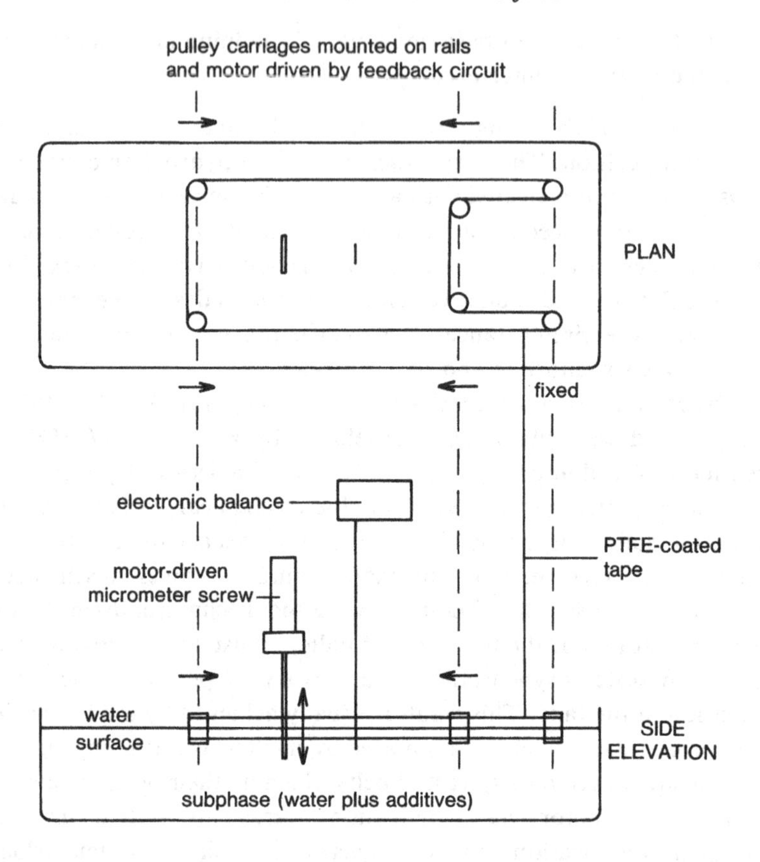

Figure 3.1. Constant perimeter Langmuir–Blodgett trough. The pulleys round which the PTFE coated tape passes are mounted on carriages which move so that the perimeter of the enclosed area remains constant but the area enclosed by the tape varies.

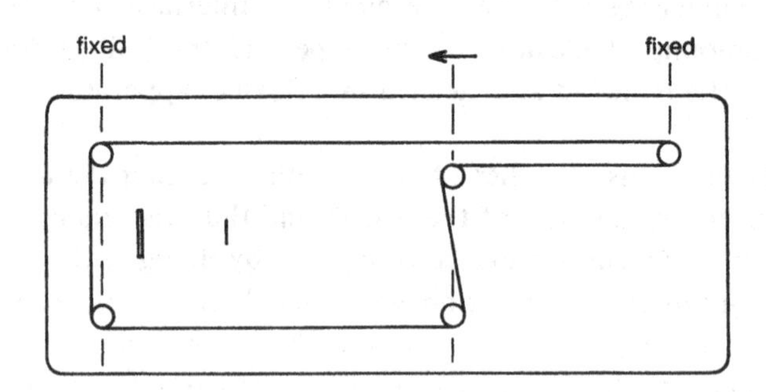

Figure 3.2. A simplified variant of the system shown in Figure 3.1.

ribbons can be pulled past the ends of a fixed barrier so as to compress the enclosed film. There are clearly defects in this simple system but it might well be developed into something more satisfactory. For materials which produce films having a high viscosity, there are always likely to be problems in producing a trough which ensures uniform compression over an entire film.

The surface pressure of a film, π, is defined as the difference between the surface tension of clean water and the surface tension of the water with the film upon it. The most popular way of measuring this quantity is the Wilhelmy plate. This consists of a small vertical rectangular plate made of a material chosen so that the contact angle of the water and superincumbent film is zero. The downward force on such a film arising from surface tension is then equal to twice the horizontal length of the plate times the surface tension. The most common choice for the material of the plate is filter paper. It is usual to suspend the plate from an electronic balance. Another method of measuring π was introduced by Malcolm and Davies [64] and has advantages when highly viscous films are studied. A thin phosphor bronze ribbon intercepts the water surface and is arranged so that the film is on one side and clean water is on the other. The ribbon exerts force on a pressure transducer which generates an electrical potential proportional to the pressure.

It is often important to control the temperature of the subphase at a point above or below ambient. This can be achieved either by a water jacket surrounding the trough or by immersing a zig-zag shaped piece of glass tube in the subphase and passing either heated or cooled water through it.

Much attention is given in the literature to the question of cleanliness and the purity of starting materials. The importance of these precautions varies depending on the work to be undertaken but the use of doubly distilled de-ionised water as a subphase has become a standard precaution. The use of disposable surgical gloves when handling pieces of apparatus which will have contact with the subphase is also a wise precaution.

It is important to mount the glove box on some form of anti-vibration device but this system need not be particularly sophisticated.

3.2 Characterisation of monolayers: classical methods

The expression classical methods is used here to describe methods which were pioneered before the Second World War and which determine the macroscopic properties of monolayers at the air/water interface. It is only by indirect inference that we can use these properties to find the microscopic structure of the film. Most of the work carried out during this pioneering period was devoted to the study of long-chain fatty acids and the vocabulary of the subject thus tends to be dominated by the ideas developed during this period. It should be understood that much of the recent work on films at the air/water interface which appears in the literature concerns other materials. However, it is only in the case of fatty acids and, to some extent, phospholipids, that any proper theoretical rationale for the observed behaviour has been developed. Most films of amphiphilic materials at the water surface are far from thermodynamic equilibrium and the application of classical thermodynamics to their behaviour must thus be used with caution. A typical film is the two-dimensional analogue of a piece of a cast solid which is polycrystalline and contains various lattice defects. In certain cases it is best described as a two-dimensional liquid crystal.

The basic characteristic of films at the water surface is the isotherm which is a plot of the surface pressure, π, versus the area per molecule and is thus the two-dimensional analogue of a normal three-dimensional isothermal plot of pressure versus volume. The area per molecule can be calculated from the concentration of the spreading solution, the quantity spread and the area of the confined film. Methods of measuring π have already been discussed. Typical isotherms are shown in Figure 3.3. With many materials there is an initial hysteresis in the isotherm and in some cases hysteresis persists indefinitely. It is a mistake to believe that the isotherm always reflects the microscopic behaviour of the material. Some materials initially form islands of molecules which, on compression, are forced together and deform. When the pressure is reduced the islands separate again. For some materials such as porphyrins these islands are even observable with the naked eye under suitable lighting. If amphiphilic materials are to form reasonably stable films on the water surface, their hydrophobic component must be large enough so that their solubility in water is low. However, it should be realised that *any* such material is soluble to a certain extent. Furthermore, the volume of the subphase is probably 10^7 times larger than the volume of the surface layer. Thus, for a high surface pressure (say 50 mN m^{-1}), equilibrium may correspond to the amphiphilic material being largely in solution in the subphase. It is thus to be expected that, under these circumstances, the film will gradually 'collapse' and disappear with the passage of time. Such a collapse should not be confused with the rapid collapse which

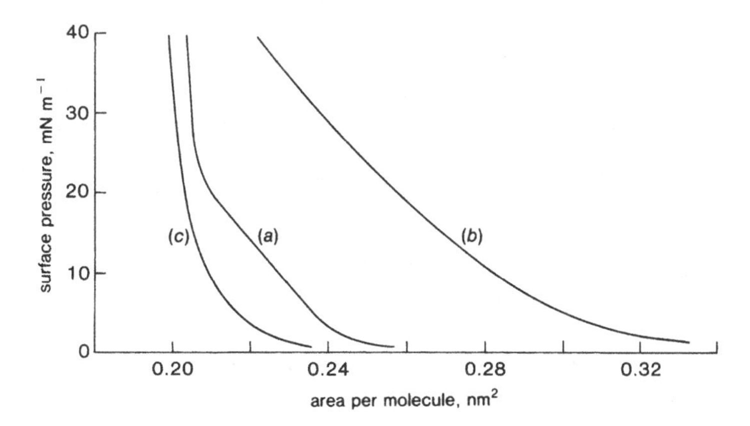

Figure 3.3. Forms of isotherm which often appear. (*a*) This is a typical isotherm observed with an impure long-chain carboxylic acid. (*b*) Isotherms such as this usually correspond to substantial changes of the conformation of the material as the pressure, π, is increased. Such materials usually do not lend themselves to the formation of Langmuir–Blodgett films. (*c*) This form of isotherm occurs when the material under study tends to form islands before compression.

arises at high pressures when pieces of film ride up and overlay other pieces, leading to the formation of regions of disordered multilayers.

Surface viscosity has an important influence on the deformation of films and can also provide information about structure. Gaines [14] describes various methods of measuring this quantity. The damped torsion pendulum as developed by Langmuir and Schaefer [65] is probably the best device for making such measurements. Recent measurements of this type have been made by Buhaenko *et al.* [66]. Malcolm [67, 68] and Daniel and Hart [69] have carried out experiments which illustrate the important influence which viscosity has on the study of isotherms.

The influence of the presence of an organic monolayer on the surface potential measured at the water surface can also provide interesting information. However, this information can be difficult to interpret as it depends on the dipole moments associated with both the upper and the lower ends of the molecules and also the surface double layer in the water immediately below the film. There are two different ways of measuring this quantity. Both, of course, involve a counter electrode below the water surface.

1. In the Kelvin method, a planar electrode is vibrated up and down just above the surface. This electrode and the surface form the two plates

of a condenser which is connected across the input terminals of a phase sensitive amplifier gated by an alternating potential derived from the vibrator. This condenser is also connected via a very high resistance to a bias voltage. If the mean potential between the condenser plates is not zero, the charge on the condenser remains approximately constant and, as the capacity varies, an alternating potential is generated and amplified. If the bias is adjusted so that the charge on the condenser is zero, no alternating output is generated. Changes in bias needed to minimise the alternating output thus correspond to changes in surface potential. The Kelvin method is difficult to apply to films at the water surface as any disturbance varies the effective capacity of the condenser formed by the top electrode and the film surface and hence generates a voltage. It is difficult to avoid a mechanical disturbance at the frequency of the vibrator and this will, of course, break through the otherwise selective phase sensitive amplifier. In any case, very careful screening is needed if this method is to work properly.

2. A more satisfactory procedure involves a radioactive probe held a little above the water surface. The radioactive radiations ionise the air between the probe and the surface and ensure that they are at the same potential. The probe is connected to an electrometer having a very high input impedance which reads the surface potential. This procedure was pioneered by Guyot [70] and Frumkin [71] in the 1920s but has become much more convenient to use with the introduction of artificial radioactive isotopes and modern electronics. Americium 241 is particularly useful as the low energies of the α and γ radiations produced only ionise the air in the immediate vicinity of the probe.

The other methods of characterising monolayers discussed by Gaines [14] are really only effective when the layer is deposited on a solid substrate or have only come into their own with improvements of technology and are thus discussed in the next chapter.

Efforts to interpret the forms of isotherms in terms of behaviour at molecular level have tended to concentrate on fatty acids and, to a lesser extent, phospholipids. An understanding of the behaviour of films at the air/water interface in terms of the degree of order in the structure of the film is thus largely limited to these materials. Harkins carried out important early work in the detailed study of the isotherms of fatty acids and this work is summarised in his book *The Physical Chemistry of Surface films* [72]. The nomenclature that he introduced was derived from an attempt to find an analogy with the behaviour of three-dimensional materials. Thus, below a critical temperature, one would expect to find a gaseous, a liquid and a solid phase separated by first order phase changes. In a first order phase transition, the molar entropy, molar volume and other molar parameters suffer discontinuous changes. The Gibbs free energy, $G = U - TS + PV$, is continuous but its partial derivatives with respect to T and P (the other quantities being treated as functions of these quantities) are

$$\left(\frac{\partial G}{\partial T}\right)_P = -S, \qquad \left(\frac{\partial G}{\partial P}\right)_T = V \tag{3.1}$$

and are discontinuous. (Here U is the internal energy, T the absolute temperature, S the entropy, P the pressure and V the volume). This formalism can be applied to a two-dimensional system if P is replaced by π and V by A. In contrast a second order phase transition is one in which the Gibbs free energy and its first derivatives are continuous but its second derivatives are discontinuous. Examples of second order transitions are the phase changes associated with the Curie temperatures in ferromagnetic materials and the order/disorder transitions observed in some alloys.

For most fatty acids, Harkins observed not three but four distinct phases which he named gaseous, liquid expanded, liquid condensed and solid phases. A schematic outline of this behaviour is shown in Figure 3.6 later. The most distinctive transition is that between the liquid expanded and the liquid condensed phases, to which there is no analogous transition in three-dimensional systems. There has been much controversy over the nature of this transition. If it was first order one would expect a plateau of constant π to appear in the isotherm, corresponding to a region in which the two phases are in equilibrium. Until fairly recently it appeared that there existed a kink but no plateau at the point on the isotherm corresponding to this transition and it was thus thought to be second order. However, Pallas and Pethica [73] showed that, in the cases of pentadecanoic acid and hexadecanoic acid (see Figure 3.4), there exists a distinct plateau which is obscured if impure material is studied. This work was carried out using an acid subphase so that the carboxylic acids would not be ionised. Subsequently a number of other workers have shown that the liquid expanded/liquid condensed transitions associated with other carboxylic acids are first order. In some cases these authors have made use of the direct study of isotherms, while other workers have made use of the study of other physical properties

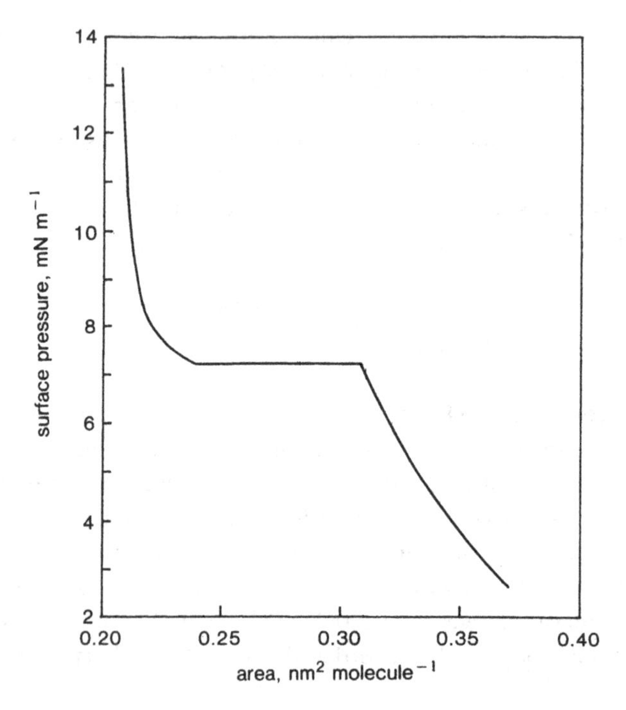

Figure 3.4. Isotherm for pentadecanoic acid at 25°C. (Taken from Pallas, N.R. and Pethica, B.A. 1985 *Langmuir* **1** 509–13. Published by permission of The American Chemical Society and the authors.) The curve follows the experimental points (not shown) closely. The two-phase region in which the liquid expanded and liquid condensed phases coexist corresponds to the plateau.

of the monolayer that demonstrate that the transition is first order. For example Winch and Earnshaw [74] studied light scattering from the thermally excited ripples at the surface.

Moore *et al.* [75] have carried out studies of pentadecanoic acid and have confirmed the results of Pallas and Pethica [73]. They have also used the fluorescent probe technique (discussed in Section 3.3) to examine this material. These authors have also confirmed the empirical rule put forward by Gaines [14] that, in a homologous series of long-chain fatty acids, there is an 8–10 K shift in the phase diagram for each CH_2 group. (This paper will be returned to in the next section.) Now, as one is studying a system at the air/water interface, the range of temperatures accessible to experiment is limited and these shifts can move phase transitions in particular cases above or below this range, so that the apparent behaviour of materials having different chain lengths is very different though their fundamental behaviour is very similar.

Rettig and Kuschel [76] have shown that the phase behaviour of the methyl esters of fatty acids is similar to that of the un-ionised acids.

In the discussion of fatty acids given so far it has been assumed that the subphase is sufficiently acid so that these materials are not ionised. However, if the pH is increased so as to be comparable to or larger than the pK_a of the fatty acid, then the situation will obviously be different. If now a divalent or trivalent cation is introduced into the subphase, the structure of the film and the resulting isotherms will be substantially changed. Wostenholme and Schulman [77] were amongst the first workers to make a systematic study of this effect. Binks [78] has given a recent review of the very extensive literature devoted to this phenomenon. The matter will be returned to in Section 3.3.

An extensive discussion of viscosity measurements will not be given here, though it should be noted that, where they have been made, there are discontinuous changes in viscosity corresponding to the various phase changes.

3.3. Characterisation of monolayers: recent techniques

The last decade has seen the introduction of several new characterisation techniques which have been of major assistance in understanding the structure of monolayers at a molecular level. The most important of these has been the use of synchrotron radiation to obtain diffraction patterns from films at the air/water surface. In principle it would always have been possible to use X-rays for this purpose but the high intensity and highly monochromatic nature of the radiation from a synchrotron source has made this technique far easier to use. A selection of recent papers based on this technique is given [79–88], not all of which refer to simple fatty acids. The information available from such experiments is of two distinct kinds, though, in several studies, both kinds of information have been obtained.

1. Monochromatic radiation impinges on the water surface at an angle, α_1, to this surface and radiation which undergoes specular reflection leaves at an angle, α_2, where α_1 and α_2 are adjusted to be equal. The intensity of the reflected radiation as a function of α gives information about the variation of the electron density of the system in the direction normal to the surface.

2. Monochromatic radiation is again directed at the water surface at a small angle, α, less than the critical angle for total external reflection. The intensity of the reflected radiation is explored as a function of the

azimuthal angle, 2θ, in a plane parallel to the water surface. This experiment is the analogue of the powder pattern experiment in three-dimensional X-ray work. For a given θ, only those regions of the film having molecular lines (the analogue of planes) making an azimuthal angle θ to both the incoming rays and the direction of the axis of the collimator associated with the detector will be detected. Of course, as in all diffraction from crystals, a limited region of coherence leads to a peak whose width is inversely proportional to the coherence length. This important technique was pioneered by Kjaer *et al.* [79] and by Dutta *et al.* [81]. It has been further studied by both groups of authors [83, 86] and we refer here particularly to the study of arachidic acid, $CH_3(CH_2)_{18}COOH$, by the former group [83]. For a subphase having pH=5.5 and values of π between 1 and 25.6 mN m^{-1} they find an ordered phase in which the area per molecule varies between 0.198 nm^2 and 0.237 nm^2 and the tilt angle varies between 0° and 33°. They deduce the tilt angle from the change in film thickness found from measurements of the vertical variation of electron density. In this phase there is a correlation length involving distances of about 15 nm, though the angular correlation may extend over greater distances. These results are consistent with a pseudo-hexagonal structure which becomes truly hexagonal when the tilt angle is zero. From a strict crystallographic point of view the structure for zero tilt is centred orthorhombic. There remains the question of whether the molecules tilt towards the nearest neighbours or the next nearest neighbours, the two arrangements being illustrated in Figure 3.5. A more detailed discussion than can be given here, making use of a comparison of predicted and measured variation of peak intensities with tilt angle, shows that the tilt is in fact towards the *nearest* neighbours.

Bohanon *et al.* [86] studied heneicosanoic acid (which contains 21 carbon atoms) and Lin *et al.* [87] studied this material with particular reference to the effect of pH and the presence of divalent cations in the subphase. The former authors made use of in-plane diffraction (method 2 above) and obtained first order *and* second order diffraction peaks. They were able to show that, at high pressures (π=35 mN m^{-1}), at low pH (pH=2) and at temperatures in the region of 0–5 °C, the material packs into a distorted hexagonal structure with the tilt towards the nearest neighbours. However, in the region 5–10°C the tilt is towards the next nearest neighbours. In the latter study [87] in-plane diffraction was studied as a function of pH and the presence of Ca^{2+} or Cu^{2+} in

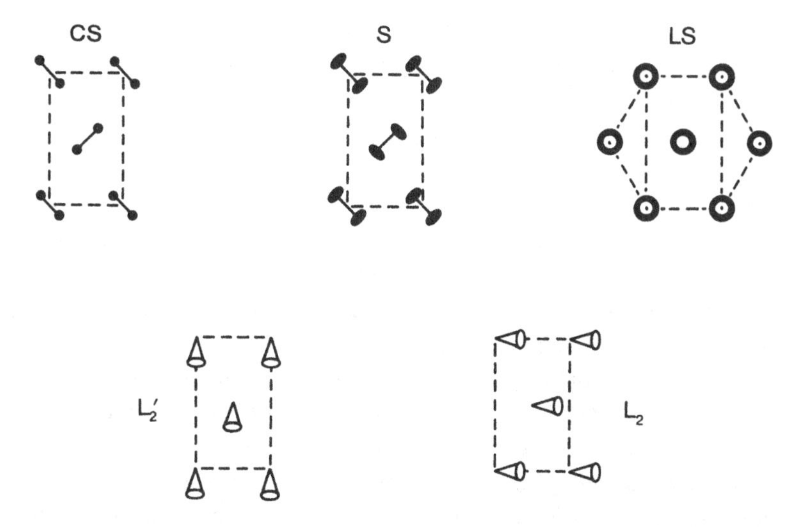

Figure 3.5. Schematic arrangement of the various phases of behenic acid (docosanoic acid) at the air/water interface corresponding to the phase diagram shown in Figure 3.6. (Taken from Kenn, R.M., Bohm, C., Bibo, A.M., Peterson, I.R., Möhwald, H., Als-Nielsen, J. and Kjaer, K. 1991 *J. Phys. Chem.* **95** 2092–7. Published by permission of the American Chemical Society and the authors.) All the phase structures are distorted forms of hexagonal packing. ⟍ denotes an end view of a molecule which stands vertically but does not rotate about its axis and is in the phase denoted by CS. ⟋ is similar to the above but is in the S phase and librates. o denotes a molecule the axis of which is vertical and which rotates about its axis. Δ denotes a tilted molecule. In the liquid expanded, L_2, phase the tilt is towards the nearest neighbour and in the liquid condensed, L_2', phase the tilt is towards the next nearest neighbour.

the subphase. The structural changes resulting from a divalent cation were not analysed in detail. However, as would be expected, the effect of these ions on structure is much more pronounced at high pH than at low pH. Indeed, at pH = 2 these ions appear to have no effect on structure.

Lin *et al.* [84] studied the long-chain alcohol heneicosanol ($C_{21}H_{43}OH$) with particular reference to the dynamics of phase change. They used a subphase of pH 2 (though it is not clear why such an extreme pH was used when an alcohol was being studied) and examined layers at $\pi = 15$, 20 and 25 mN m^{-1} at temperatures ranging from 5°C to 30°C. They assumed that diffraction patterns involving two pairs of peaks correspond to a tilted hexagonal structure and that a single pair of peaks correspond to an untilted structure. Above 20°C they observed only a hexagonal structure while at 5°C they observed only a tilted hexagonal structure. In the range 10°C to 20°C it is possible to observe slow changes of phase with time, involving time constants of the order 10^3 s.

Recently Kenn *et al.* [89] have studied films of docosanoic acid and have used their results to construct a phase diagram for this material which they believe to be representative for most fatty acids. This work will be discussed in the next section.

The other important technique for the study of films at the air/water interface which has recently been introduced is fluorescent microscopy. This technique was introduced by von Tscharner and McConnell [90] and Möhwald [91, 92]. It depends on the fact that certain amphiphilic fluorescent dyes become incorporated into islands of the surface active material under study. Furthermore, where two phases of the surface active material coexist, the dye can often be chosen so that it segregates preferentially into one phase. A shallow Teflon trough is employed with a water immersion objective incorporated into the bottom. The depth of water is adjusted so that the objective focuses on the water surface. The layer of material at the air/water interface is illuminated by a xenon lamp. The fluorescent light so generated passes via the objective and suitable filters to an image-intensified video camera and the image is displayed on a television screen. In some versions of this technique the fluoresence is viewed from above. Most of the pioneering work in this field was devoted to the study of phospholipids, a topic to which we will return. Recently this technique has been applied to the study of pentadecanoic acid and this work will be considered here as it relates directly to other papers discussed in this section.

Moore *et al.* [75] made an extensive study of pentadecanoic acid. They employed the fluorescent probe, NBD-HDA, 4-(hexadecylamino)-7-nitrobenzo-2-oxa-1,3-diazole. They explored the liquid-expanded–gas (LE–G) and liquid-condensed–liquid-expanded (LC–LE) transitions. Measurements were made at a series of temperatures. For example, at 25°C with a molecular area of 0.61 nm^2, which is in the region where the LE and G regions coexist, they observed a bright field containing dark circular regions. The dark regions are bubbles of two-dimensional gas surrounded by the liquid phase. The contrast between the two phases is the result both of the difference in density of material, and hence of dye, and of the quenching of the dye in the gas phase. Isothermal compression of the film leads to disappearance of the gas bubbles at an area of 0.43 nm^2, which result determines the boundary between the LE–G coexistence region and the LE region at this temperature. Further compression produces a homogeneous film in the region 0.43 nm^2 to

0.32 nm^2. Below 0.32 nm^2 the image is once again heterogeneous, the dark circular regions now being associated with the liquid condensed phase. The contrast is now due to the fact that the dye is less soluble in the more condensed phase. Further compression leads to a disappearance of the LE phase at an area of about 0.2 nm^2 per molecule. Clearly repetition of this experiment at different temperatures makes it possible to establish the phase boundaries on the temperature–area plot with considerable precision. The results of this study confirm the findings of Pallas and Pethica [73] to within an acceptably small error.

Monolayers of pentadecanoic acid at the air/water interface have been studied intensively in the region of LE–G coexistence by two different groups using the fluorescent probe technique. Stine *et al.* [93] and Berge *et al.* [94] have studied the kinetics of bubble evolution and find, for large gas coverage, that the bubbles take on a polygonal structure and that the mean bubble area grows with time according to theoretical predictions already in the literature.

Much of the work carried out using the fluorescent technique has been devoted to lipids. These materials will be discussed in Chapter 8. For our present purposes it is sufficient to think of lipids as being amphiphilic molecules consisting of two parallel hydrophobic hydrocarbon chains and a hydrophilic head group. There exist many different lipids distinguished both by the lengths of the hydrophobic chains and by the nature of the hydrophilic head group. The majority of these occur naturally in cell membranes. Möhwald and his collaborators [92, 95–9] have made an extensive study of lipids at the air/water interface using fluorescent probes. Their studies have been devoted to the region in which an apparently solid or gel-like phase coexists with a fluid phase. Also, at very low surface pressure, there is evidence for a transition to a gaseous phase. The gel-like regions initially take on fractal configurations (See Mandelbrot's book [100] for a simple explanation of fractals.) The equilibrium forms which these regions take on, however, depend on achieving configurations which produce a local minimum in the free energy. The problem has been addressed by McConnell and his collaborators [101–3]. Two types of force are involved. On the one hand, the vertical electric dipoles associated with the lipid molecules contribute a repulsive effect, but on the other, the line tension associated with the boundary of a region tends to decrease its perimeter. Vanderlick and Möhwald [99] have examined the predictions of this theory and compared these predictions with experimental results. The theory predicts gel regions consisting of a regular series of lobes distributed around central roughly circular regions. Such predictions correspond approximately with the observed structure of regions and the reader is referred to this paper for further discussion of this topic.

The generation and detection of second harmonic radiation produced by irradiating a monolayer at the air/water interface by a powerful laser beam is another recently introduced technique for the study of such layers. A second harmonic can only be produced from a non-centrosymmetric target and clearly the interface between two phases is non-centrosymmetric. If a monolayer of an amphiphilic material is spread on the water there will be a change in the non-linear effect. Furthermore, a fundamental wave having an electric vector approximately normal to the water surface can produce second harmonic components having the electric vector both normal to and tangential to the water surface. It is possible to use knowledge of the relative intensities of these components to deduce the mean tilt angle of those parts of the amphiphilic material which make the main contribution to second harmonic generation. This technique has been developed and employed by Shen and his collaborators [104–7]. It has also been employed by Kajikawa *et al.* [108], Shirota *et al.* [109], Vogel *et al.* [110] and Zhao *et al.* [111].

In general the results obtained confirm and supplement the results arising from the techniques discussed in the early part of this section.

3.4 Monolayers of straight chain compounds: theoretical ideas

In order to bring together the empirical data presented above and to attempt to provide a theoretical rationale for it we turn initially to the case of docosanoic acid, $CH_3(CH_2)_{20}COOH$. This material was the subject of an extensive study by Stallberg-Stenhagen and Stenhagen [112] and has the advantage that the chain length is such that most possible phase changes are accessible to experiment. It is for this reason that Kenn *et al.* [89] chose this material for an extensive study by synchrotron radiation diffraction in the film plane. Figure 3.6 is a phase diagram for the material where the symbols for the different phases are taken from [112].

CS denotes the close packed solid phase.
S denotes the high pressure solid phase.
LS denotes a low viscosity liquid phase.
L_2' denotes a liquid compressed phase.
L_2 denotes a liquid expanded phase.

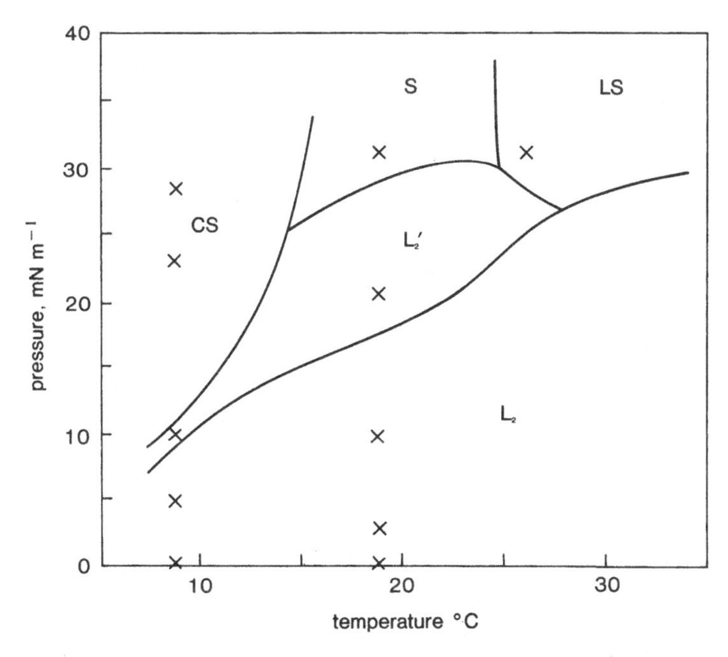

Figure 3.6. Phase diagram for behenic acid at the air/water interface. The symbol × denotes the points at which synchrotron radiation diffraction experiments were carried out. For the notation and acknowledgements see Figure 3.5.

This diagram is based on a study of the isotherms obtained from the study of this material. The symbol × denotes points at which synchrotron radiation diffraction experiments were carried out. From these measurements it is possible to deduce that the various phases have the structures given in Table 3.1.

The correlation lengths deduced from peak widths given in lattice spacings are about 150 nm for CS whereas the mean correlation lengths for the other four phases are less than 45 nm. In phases S, L_2, and L_2' the positional correlation length is up to five times smaller along the tilt direction than in the direction perpendicular to it. It seems likely that the other un-ionised fatty acids behave in a similar way, but materials with shorter chains have phase changes at temperatures lower by about 8°C per CH_2 unit and thus the regions equivalent to the lower parts of Figure 3.6 are not accessible.

It is clear that the many phases shown in Figure 3.6 have no simple equivalence to the three-dimensional phases solid, liquid and gas. However, it is well known that liquid crystalline materials exhibit a wide range

Table 3.1 *Structures of the various phases.*

Phase structure		Equivalent smectic liquid crystal phase
CS	Solid	–
S	Orthorhombic with no rotation about the long axis	E
L_2	Tilted hexagonal, tilt to nearest neighbours	I
L_2'	Tilted hexagonal, tilt to next nearest neighbours	F
LS	Hexagonal, rotation about axis	B

of distinct phases. Bibo *et al.* [113] have suggested that the phases exhibited by straight chain amphiphiles at the air/water interface correspond very closely to the various smectic mesophases observed in three-dimensional systems. There is, of course, the distinction that, in three-dimensional systems, degree of correlation in the position of molecules between molecular planes as well as within planes is important. The identification of monolayer phases with liquid crystalline phases in the table above seems totally convincing. Bibo *et al.* [113] have argued that other phases appear in the monolayer and that these can be related to other smectic liquid crystal structures. However, these authors accept that their provisional findings need to be confirmed by other techniques.

Nelson and Halperin [114] have considered the theory of melting in a two-dimensional lattice in which the lowest energy state corresponds to a population of 'point' molecules situated on a triangular (or two-dimensional close packed hexagonal) lattice. This highly mathematical paper has been ably summarised by Brock *et al.* [115]. The predictions made by Nelson and Halperin [114] can be summarised as follows. For a two-dimensional material having a close packed hexagonal structure one can, by definition, expect both the positional and the orientational order to be long range. However, at a certain temperature, T_m, the long range positional order disappears and the orientational order also ceases to be truly long range but decreases according to an algebraic law rather than according to an exponential law. Thus regions of material will exist in which the orientational order appears to be similar to that existing in a two-dimensional solid but in which positional order is short range. Such a structure is said to be hexatic.

In order to discuss the hexatic phase it is necessary to introduce the idea of a disclination. Imagine a two-dimensional close packed hexagonal lattice drawn on a deformable sheet. If one chooses a particular lattice site as the centre of coordinates, the lattice will consist of six 60° sectors centred on this point. One now has two alternatives.

1. Make a cut along one lattice line which passes through the origin and also along a lattice line located at plus or minus 60° from the original line and also passing through the origin. One can now remove the 60° sector so produced and then deform the sheet so as to close up the gap left by its removal. This deformation is equivalent to the introduction of a series of dislocations.
2. Make a cut along one lattice line and then deform the sheet so as to accommodate an additional 60° sector.

The two structures so formed are analogous to a particle and its antiparticle. They can annihilate one another with release of energy but otherwise can only be destroyed at the boundary of the medium. In a similar manner, a suitable injection of energy can create a 'disclination pair' in the body of the material.

The hexatic phase contains disclination pairs which are attracted to one another and thus tend to produce bound pairs. Nelson and Halperin [114] showed that, at a temperature which they denote by T_h, the disclination pairs become unbound, and that this process leads to melting and the formation of a true two-dimensional liquid. In those phases which are observed at the air/water interface and where a sixfold coordination exists, it is possible also for a uniform tilt to exist over substantial regions. In this case the situation is a little more complex. Such phases do not have sixfold symmetry, so there is not a natural seamless join along the cuts envisaged in the processes 1 and 2 described above. Thus a line analogous to a twin boundary will occur along the cut. Such a line must originate at a disclination of one sign and terminate at a disclination of the opposite sign, though it need not be a straight line. This interesting idea has been pursued by Peterson who has found evidence supporting it in the study of Langmuir–Blodgett films of ω-tricosenoic acid and this topic will be returned to in Chapter 4.

Apart from the case of straight chain carboxylic acid films and films formed from closely related materials, there exists only a very limited amount of evidence bearing on the detailed structure of films at the air/water interface. Most of the available evidence is indirect and depends on the study of Langmuir–Blodgett films. As there is always the possibility of rearrangement during the process of deposition of these films, such evidence must always be treated with caution.

In Chapter 5 we will discuss the formation of LB films made from rigid-rod polymers. Prior to deposition, these materials lie with the polymer chain axis parallel to the water surface and they thus constitute a two-dimensional analogue of a nematic liquid crystal. One would thus expect a local anisotropy of physical properties which, if it could be made to extend over macroscopic distances, would lead to anisotropic properties which could be observed experimentally. Tredgold and Jones [118] studied the velocity of ripples at a water surface bearing monolayers or bilayers of poly(γ-benzyl-L-glutamate) or poly(γ-methyl-L-glutamate) and showed that the compression process leads to a macroscopic anisotropy of surface tension which was stable for several minutes. A vibrating probe generated a ripple pattern which was circular for ordinary materials but elliptical when an anisotropic material was studied. The pattern could be viewed and photographed using stroboscopic lighting.

4
Langmuir–Blodgett films

4.1 The formation of Langmuir–Blodgett films

The technology of the formation of LB films has much in common with the trough technology discussed in Chapter 3. In addition to the equipment described there, one needs a mechanical device to raise and lower the substrate through the air/water interface at a predetermined rate. Various devices have been employed but it is usual to provide the vertical movement by driving a large micrometer screw by an electric motor via a reduction gear train. Practical velocities are such that they are usually measured in millimetres per minute. It is essential to be able to vary the dipping rate as there is an upper effective rate which can be employed for any particular material. This rate is determined by the speed at which water drains from the film as it is withdrawn from the subphase and by the viscosity of the film and hence the rate at which material can approach the substrate. For a material of high viscosity this procedure is difficult to carry out properly and a substantial difference in pressure may occur between the pressure sensor and the region immediately in contact with the substrate. In fully automated troughs the substrate is withdrawn from the subphase and maintained in this position while the film at the air/water interface is respread and compressed to a predetermined dipping pressure. In such systems it is necessary to carry through a cyclic compression and expansion process several times to arrive at a good approximation to an equilibrium situation before the dipping process is reactivated.

It will be obvious that, when the substrate is much smaller than the trough and the surface viscosity is appreciable, there will exist a complicated flow pattern round the substrate which may cause difficulties. Various ingenious devices designed to overcome this problem have been described in the literature but the problem has still not been totally solved.

The actual deposition process can be visualised as taking place in the manner shown in Figure 4.1. Here it is supposed that, on the upward stroke, hydrophilic interaction is responsible for adhesion and that, on the downward stroke, hydrophobic interaction is responsible for adhesion. The deposition ratio is defined as the ratio of the area of film deposited to the change in area at the air/water interface corresponding to this deposition. If the deposition is very near unity, it is assumed that perfect deposition has taken place. If the deposition ratio is near unity for both upward and downward strokes, the material is said to be deposited in the Y mode. If this ratio is near unity on the up stroke and near zero on the down stroke, the deposition is said to be in the Z mode and the converse situation is said to lead to deposition in the X mode. A simplistic interpretation of Z and X deposition would lead one to suppose that such deposition would lead to a non-centrosymmetric structure. In some cases this is indeed true but, in many cases, there appears to be some kind of rearrangement after deposition so that apparent X or Z deposition leads to a structure similar to that which one would achieve with Y deposition and a structure of regular bilayers is produced.

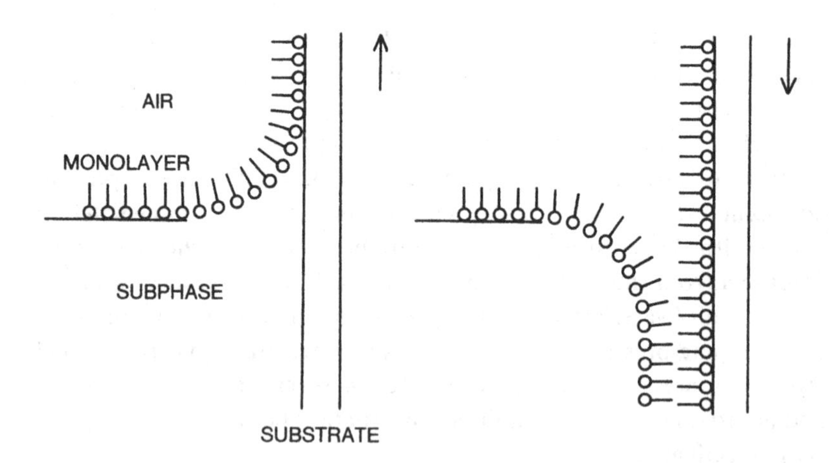

Figure 4.1. Schematic diagram of the formation of a Langmuir–Blodgett film. Each amphiphilic molecule is represented by a circle with a tail, where the circle denotes the hydrophilic end of the molecule. The left hand diagram represents the deposition of a monolayer on a hydrophilic substrate moving upwards. The right hand diagram represents the deposition of a second layer during the downward movement of the substrate.

It is important at this point to consider the nature of the substrate and its influence on the film. It must be remembered that a monolayer is only a few nanometres thick and thus even very small irregularities can have a severe influence on the final LB film. For this reason a mechanical polish which leaves micro-scratches is useless. Metal layers deposited by evaporation *in vacuo* tend to form tiny granules which, though invisible under the optical microscope, are large compared with a monolayer thickness. Furthermore, with the exception of the noble metals, most metals rapidly oxidise in air or water and form granular oxide layers. Aluminium appears to be an exception to these generalisations and is discussed in the next section. It is common to use a hydrophilic substrate; glassy oxides and glassy chalcogenides are useful in this respect. Flame annealing or other high temperature treatment appears to be capable of forming a surface which is very nearly atomically flat. Such surfaces when freshly etched to remove air borne contaminants, are capable of forming very good hydrophilic substrates. Clearly in such cases the first layer is deposited on the upstroke. If a hydrophobic substrate is required, it is possible to treat a glass surface in a hexamethyldisilazane vapour for several hours. A small quantity of this material placed in a closed glass vessel with the substrate at room temperature is sufficient.

Most semiconductor surfaces are reactive and form oxide surfaces in air or water and thus suffer from the same faults as metals in this respect. This is not true of the layer lattice materials which form valence bonds only in the layer plane. Freshly cleaved surfaces of pyrolytic graphite or molybdenum disulphide are thus promising materials, though they are only available in small areas. In principle it might be possible to form LB layers on these materials by epitaxy but very little progress has so far been made in this direction.

4.2 Simple Langmuir–Blodgett films formed from carboxylic acids

Blodgett's original paper [13] on LB films was devoted to systems formed from fatty acids and there has been a strong tendency to return to the study of these materials, particularly by those workers who are interested in the basic physics of the subject. It is natural therefore that a discussion of LB films should start with these materials. The literature of this topic is very extensive and the reader is referred to the recent book edited by Roberts [119] for a discussion of earlier papers in this field. In Section 2.4 the study of these materials by X-ray diffraction was discussed. Such studies establish the fact that the order of LB films in the direction normal to the substrate and of the lattice planes is extremely good and it was precisely this order which was initially responsible for an enthusiasm for the study of LB films. It was assumed that monolayers could be used as gate electrodes in field effect transistors and would even be of use in devices depending on electron tunnelling through monolayers of long-chain carboxylic acids. Handy and Scala [120] published an important early paper in which some of the problems inherent in such experiments are discussed. Most subsequent workers in this field have used aluminium evaporated *in vacuo* on to a microscope slide as the lower electrode and Mann and Kuhn [121] and other members of this school [122–5] claimed to observe quantum mechanical tunnelling through monolayers of fatty acids directly from the aluminium to a top electrode of some other metal evaporated *in vacuo* on to the organic material. Studies made by Tredgold *et al.* [126] have shown that aluminium is unique in this respect and that monolayers of amphiphiles deposited on any other metal and covered by an evaporated top electrode always lead to a short circuit rather than apparent tunnelling phenomena. In this paper [126] it was shown that aluminium rapidly forms a closely knit oxide layer about 5 nm thick which, contrary to early ideas on the subject, remains stable at about this thickness even at temperatures as high as 200°C.

Peterson [127] made a pioneering study of conduction through monolayers of eicosanoic acid deposited on nickel and found a pattern of conductivity statistics which could best be understood in terms of conduction through defects. Procarione and Kauffman [128] showed that apparent tunnelling currents through monolayers of lipids could be varied by several orders of magnitude by an appropriate annealing process. Tredgold and Winter [129] studied stearic acid monolayers deposited on Al–Al_2O_3 and showed that, even when the possibility of growth of the aluminium oxide layer has been eliminated, annealing can reduce the apparent tunnelling current by two orders of magnitude. In this case the *shape* of the current versus voltage curves remains the same for experiments carried out at room temperature and at 4 K, though their magnitude is greatly decreased by annealing in either case. They deduced that the current was passing through similar defects but that the number of defects is much reduced by annealing. Peterson [130] succeeded in decorating localised defects with copper by electrolysis using copper sulphate and thus gave a direct demonstration of the fact that the current travels via localised defects rather than by tunnelling uniformly across all the film. Various other experiments have since been carried out in which electrolysis is used to demonstrate that the apparent tunnelling observed in organic monolayers really arises from conduction via defects. There is, of course, a distinction between experiments which employ a bottom electrode consisting of Al–Al_2O_3 and those using other metals. In the latter cases the oxide layer produced when the metal is exposed to air is rough and uneven and, when a noble metal is employed, of course, no proper oxide is formed. In the case of aluminium the oxide layer is continuous and sufficiently resistive so that the conductivity via defects is reduced to such a small magnitude that it can be confused with tunnelling.

One must now consider the nature of the localised defects, the existence of which is manifest from the data on pseudo-tunnelling discussed above. Obviously, small dust particles can be a problem here but, even when these have been eliminated, localised conduction still takes place. There exist two techniques which can give direct information about in-plane order in LB films. One of these is electron diffraction, which gives information about order on the molecular scale but is only able to examine a very small region of the film at one time. The other is optical microscopy using polarised light. As many carboxylic acid films exist in phases which involve a uniform tilt over a region large compared with the wavelength of light, they show a local birefringence which becomes observable for films consisting of more than about 50 layers. In most cases, once a two-dimensional structure has been established, it will propagate through the film as further layers are deposited, as was first shown by Peterson [131], and thus the fact that one needs of the order 50 layers to make an observation is not really a limitation. It will be seen that there exists a gap in the scale on which investigations can be made between electron diffraction and optical microscopy. To bridge this gap requires a certain amount of ingenuity.

Peterson [132] has suggested that disclination structures exist in the film existing at the air/water interface and are transferred to the substrate in the initial layer (see Section 3.4). It is then possible to develop these structures by epitaxial growth of subsequent layers and to infer the nature of the disclinations and twin boundaries joining disclination pairs by study of the spatial variation of tilt using polarising microscopy. Bibo and Peterson [133] have applied this technique to the study of ω-tricosenoic acid multilayers. They suggest that annealing at the air/water interface can greatly reduce the density of disclinations so that the subsequent multilayer will also contain a low disclination density. It is further supposed that it is the regions immediately associated with the disclinations which are responsible for electronic conduction through the film.

In order to understand papers dealing with the study of thin layers of fatty acids which make use of electron diffraction, it is necessary to understand the terminology used by workers in this field. A long hydrocarbon chain can be thought of as consisting of C_2H_4 units. If one is concerned with the packing of arrays of parallel hydrocarbon chains, a subcell is the unit cell formed from these units and is quite distinct from the main cell which characterises the packing of whole molecules. Kitaigorodskii [134] first gave a systematic discussion of this question. He pointed out that subcells can have orthorhombic, monoclinic or triclinic symmetry. The application of this concept to the stacking of multilayers of finite chains has been clearly discussed by Robinson *et al.* [135]. There turn out to be ten possible stacking arrangements, as indeed was first shown by Kitaigorodskii. However, it is important to bear in mind here that the interpretation of the birefringence patterns seen in LB multilayers of fatty acids made by Peterson [132] depends on the material being in a hexatic-like phase in which there exists a local tilt. The cells, as opposed to subcells, observed in layers of carboxylic acids are all based on a hexagonal structure though this is deformed by a tilt either towards the nearest neighbours or towards the next nearest neighbours. The characteristic feature of a disclination observed in a multilayer by a polarising microscope is a line dividing two regions having different tilt directions indicated by different optical axes, the line terminating at points at which the two tilt directions merge and become one. These points correspond, of course, to a disclination and its related anti-disclination. This is in contrast to the behaviour of a polycrystalline material in which such grain boundaries always exist as a pattern of closed loops. Furthermore, the tilt orientation can change within one domain, in contrast to a conventional polycrystalline material, in which the orientation remains constant throughout a crystallite.

Recent relevant work has made use of transmission electron diffraction, the theory of which we discussed in Section 2.5. This technique has a very straightforward interpretation which, combined with improved experimental methods, has provided very precise results. Two papers are of particular importance in this context. Garoff *et al.* [137] studied monolayers of cadmium stearate. Their substrates were 2 nm layers of SiO coated on 10 nm layers of amorphous carbon on 200 mesh Ni electron microscope grids. Each grid was mounted in a milled indentation on a silicon wafer so that the top surface of the grid was flush with the wafer surface. Deposition was made in the 'solid phase' at an area per molecule of 0.23 nm^2. An electron energy of 100 keV was employed, which corresponds to a wavelength of 0.0037 nm. They obtained hexagonal diffraction patterns in which only the first order spots had a significant intensity. Calculations indicate that each molecule occupies 0.21 nm^2 which is in good agreement with the area occupied in the initial film. Measurements were made of spot intensity as a function of both radius and angle. From these measurements it can be deduced that there exists a rapid decrease of positional correlation in the radial direction but that the angular behaviour indicates a bond orientation which remains highly correlated over distances corresponding to 10^6 unit cells. The presence of a true hexatic phase is thus demonstrated. Peterson *et al.* [138] have studied the distribution of intensity within the diffraction peaks obtained from monolayers of cadmium stearate dipped on layers of Formvar less than 20 nm thick. They compare their results with theoretical predictions calculated from three different models, one of which is based on the hexatic phase. The best fit is obtained from the hexatic model.

Bonnerot *et al.* [139] have made a study of docosanoic acid and ω-tricosenoic acid using reflection electron diffraction supplemented by infrared studies. They studied both monolayers and multilayers. Their results for monolayers agree well with the other work quoted above. On deposition of subsequent layers, a structural transition is observed which extends over the first seven layers. For thicker films the hexagonal subcell becomes orthorhombic and the main cell becomes monoclinic. (For those not familiar with crystallographic terminology, the relationship between these various structures needs explanation. Consider a two-dimensional lattice in which the individual molecular chains pack in a close packed hexagonal structure. If successive two-dimensional planes are located in register with one another and the chains are normal to the layer planes, then the main cell is a particular case of the orthorhombic structure. If the chains tilt in the direction of the next nearest neighbours, then the main cell is a particular case of the monoclinic structure.) These results for thicker films confirm the general conclusions of earlier workers and are themselves confirmed by recent work by Robinson *et al.* [140] on ω-tricosenoic acid who find either monoclinic or orthorhombic subcells existing in multilayers. This latter form of packing is known to be the equilibrium bulk form [141].

The question now arises as to what structure exists on a somewhat larger scale. If the material is in the hexatic phase, there will exist fluctuations in the distances between molecules. The resultant electron diffraction pattern will now correspond to an orthorhombic lattice but be somewhat 'fuzzed out', particularly in the radial direction. This is in fact what is observed as is shown in [138]. In contrast to these observations, a polycrystalline structure would give a diffraction pattern in which the diffraction maxima predicted for an infinite perfect lattice were replaced by clearly defined arcs if the crystallites were of a comparable size to the beam width. If the crystallites were much smaller than the beam width, the diffraction pattern would consist of rings. The pattern actually observed consists of discrete spots of finite size indicating an angular structure that changes little over the beam width, combined with a fluctuation in molecular spacing over much shorter distances. Thus there exists quite good evidence for the hypothesis that LB multilayers of carboxylic acids exist in some form of the hexatic state.

Whereas the disclination picture provides a convincing explanation of the properties of multilayers of fatty acids, being consistent with both the electron diffraction and optical evidence, it is not yet proven that, at room temperature, such systems are really in the hexatic state. It is equally probable that the existence of an initial hexatic monolayer on which subsequent layers are grown by epitaxy produces a material which is far from thermal equilibrium and has more of the nature of a glass state rather than of a mesophase. Indeed, the relative hardness of multilayers and their resemblance to true three-dimensional crystals of fatty acids tends to support this view.

One might hope that STM would shed light on the structure of monolayers. However, tunnelling over a distance as large as 2.5 nm would lead to very small currents. Recently, however, Matsuda *et al.* [142] have obtained unambiguous STM images of docosanoic acid molecules lying *on their sides*. Such images are shown in Figure 4.2. Either oriented pyrolytic graphite or small crystals of MoS_2 were used as the substrate and the film at the air/water interface was in the L_2 tilted hexagonal phase. The films were deposited either by one downward and one upward stroke or by one upward stroke made through a film which was spread after the substrate had been immersed. These authors discuss two possible mechanisms for the formation of these films but incline to the view that the films as first deposited are porous and that the STM tip removes some of the material so that the rest can rearrange into the structure observed. The two structures shown have periodicities of 0.4 nm in one

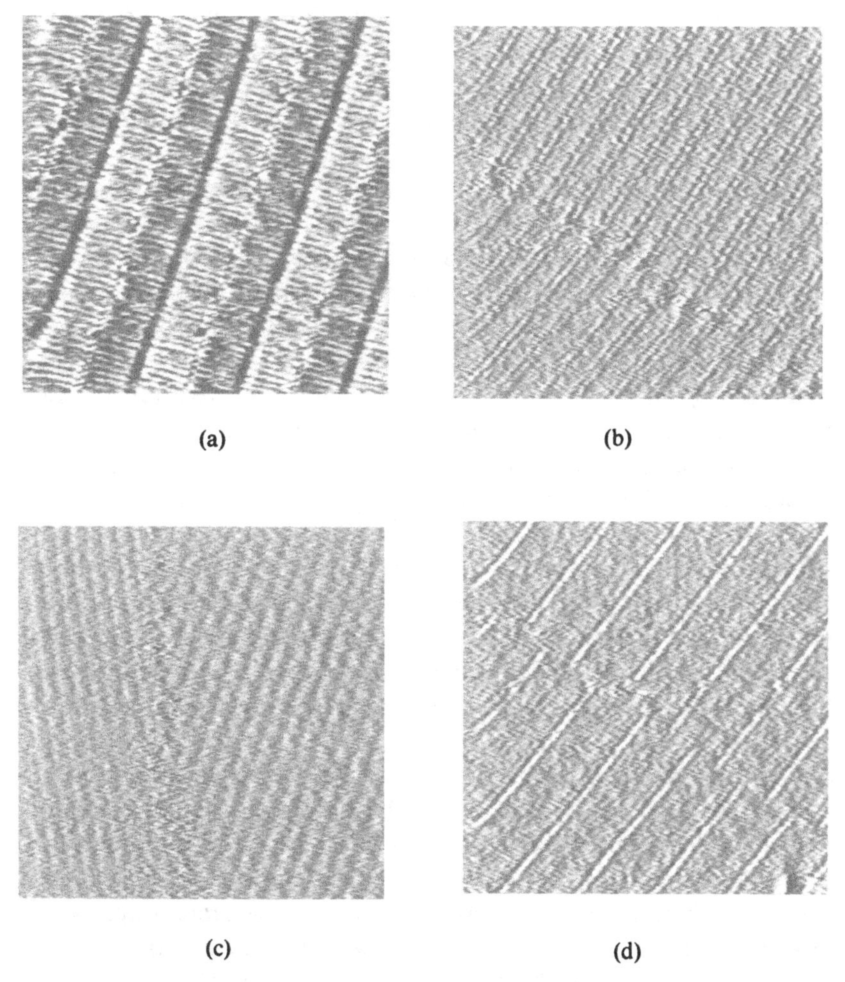

Figure 4.2. Scanning tunnelling microscopy images of prone carboxylic acid molecules (reproduced by kind permission of Dr H. Matsuda and Canon Inc.). (*a*) 22-Tricosenoic acid down-up film deposited on HOPG (pyrolytic graphite) at $\pi = 15\,\mathrm{mN\,m^{-1}}$ and 20°C (L_2 phase), scanning area $22 \times 22\,\mathrm{nm}^2$. (*b*) Docosanoic acid down-up film deposited on HOPG at $\pi = 22\,\mathrm{mN\,m^{-1}}$ and 20°C (L_2/L_2'), scanning area $36 \times 36\,\mathrm{nm}^2$. Note the domain boundary in the film where the molecules form an interdigitating structure. (*c*) Docosanoic acid one-down film deposited on HOPG at $\pi = 15\,\mathrm{mN\,m^{-1}}$ and 20°C (L_2 phase), scanning area $71 \times 71\,\mathrm{nm}^2$. Note the domain boundary in the film where the molecules form a head to head, tail to tail structure. (*d*) Docosanoic acid down-up film deposited on HOPG at $\pi = 22\,\mathrm{mN\,m^{-1}}$ and 20°C (L_2/L_2'), scanning area $36 \times 36\,\mathrm{nm}^2$. This is the one-dimensional analogue of a screw dislocation.

direction and either 3 nm or 6 nm in a direction at right angles to this. These dimensions correspond to the molecular width and either the molecular length or twice the molecular length and thus seem to be very convincing. Furthermore, tunnelling over a distance of 0.4 nm is quite possible.

Recently several different groups have made significant progress in the study of LB films by AFM. Garnaes *et al.* [463] studied cadmium arachidate deposited on 100 surfaces of silicon. The majority of their published results were obtained using four layers of this material on a hydrophobic surface prepared by a 5 s etch in 10% hydrofluoric acid. They were able to obtain atomic resolution and to observe what they claim were grain boundaries. In view of results presented earlier in this book, these boundaries seem more likely to be brought about by the presence of disclinations but the issue is still open. Chi *et al.* [464] studied LB films of several different materials using AFM. However, their most interesting results were obtained on films of cadmium arachidate consisting of either two or four layers deposited on the 100 surface of silicon rendered hydrophobic by etching in concentrated NH_4F. The two layer films exhibited an ageing process which appeared to be still going on 100 days after deposition. The four layer films did not have a uniform surface but showed a terrace-like structure, the steps between the terraces being about 5 nm in height. The horizontal extent of the terraces was of the order 0.3 μm. 5 nm is, of course, almost exactly the thickness of a bilayer of cadmium arachidate and thus it appears that, after deposition, films rearrange themselves but that the head to head bonding is sufficiently strong that the *bilayer* structure remains intact. Ali-Adib *et al.* [465] have studied multilayers of cadmium stearate deposited on hydrophilic glass. Odd numbers of layers were deposited so that the final surface should consist of hydrophobic tail groups. AFM studies showed that a surface roughness existed, which was dominated by irregularly shaped islands. The number density and average height of these islands appeared to increase as the number of deposited monolayers increased from 5 to 51. For the thinner films the general behaviour was very like that observed by Chi *et al.* [464]. These authors observed similar behaviour with various other LB materials.

The behaviour reported above has to be reconciled with the other well known properties of multilayers of cadmium stearate and cadmium arachidate. One possible picture is that these materials deposit in a hexatic structure which, for a multilayer, represents a metastable state. The upper part of the multilayer, where the molecules are free to move, tends to rearrange so as to form a number of crystallites while the lower layers remain in much the same state as when they were deposited. If the microcrystalline region is thin compared with the lower region, then the known properties of multilayers of these materials would be accounted for.

4.3 Other straight chain amphiphilic compounds

In addition to extensive studies made on carboxylic acids, numerous papers have been published which deal with straight chain materials having other hydrophilic groups at one end. Two such structures which are capable of stabilising LB films are the alcohol group and the methyl ester group, both of which are, of course, less hydrophilic than carboxylic acids and are largely unaffected by the pH of the subphase. However, very little work has been done on the characterisation of the structure of LB films formed from these materials and they are, in other respects, rather uninteresting. It is also possible to form films from molecules terminated by an amine group. Here again, behaviour is dependent on the pH of the subphase. Study of films made from amines alone has been very limited and this topic will be returned to in the next chapter when films made from alternating layers of two different compounds are discussed.

However, it is possible to introduce two new factors into the problem of multilayer structure, either of which can influence structure in an interesting way. It is possible to replace the ordinary hydrocarbon chain by a perfluorinated chain and it is possible to dip over a subphase containing a trivalent cation. The perfluorinated carbon chain has a helical structure which produces a chain which appears to be approximately cylindrical. The handedness of a chain is determined randomly so that an equal number of right handed and left handed molecules exist in a given batch.

Completely perfluorinated carboxylic acids have never been deposited with total success as LB films, as it is very difficult to synthesise materials of this nature containing more than 12 carbon atoms. Unfortunately, it is not possible to form stable monolayers at the air/water surface or dip successfully when making use of such short chains. Nakahama *et al.* [143] claim to have carried out these processes using a subphase containing aluminium ions. However, extensive efforts to reproduce these results which were made in the author's laboratory proved unsuccessful. Naselli *et al.* [144] studied the partially perfluorinated material perfluoroctylundecanoic acid ω-$F(CF_2)_8(CH_2)_{10}COOH$. Using cadmium chloride in the subphase, they were able to form stable multilayers of the cadmium salt of this material. They employed both grazing incidence reflection and transmission polarised infrared measurements and showed that the chains are substantially tilted with respect to the normal to the substrate surface. They attribute this to the mismatch in cross sectional area of the fluorocarbon and hydrocarbon parts of the molecule. As will be seen later in this book, it is possible to form Y layers of totally perfluorinated fatty acids by evaporation *in vacuo* [145] and these also involve a tilted structure, so the mismatch postulate is somewhat suspect.

Various efforts have been made to form LB films of fatty acids using a subphase containing trivalent cations. Reference has already been made to the pioneering work of Wostenholme and Schulman [77] and the review article by Binks [78]. Two recent attempts have been made to carry out such a programme by Prakash *et al.* [146] and Ohe *et al.* [147]. Both groups of workers made use of trivalent iron in the subphase and had some success, but X-ray diffraction data showed that the films obtained were far less well ordered than films dipped over divalent cations.

4.4 Azobenzenes and related compounds

Moving from simple rod-like molecules, one arrives naturally at those amphiphiles which consist of a rod-like structure containing near its centre a cyclic group. A considerable amount of research has been devoted to the study of materials containing both the azobenzene structure and the closely related stilbene structure, which are illustrated in Figure 4.3.

Chemists will be familiar with the bonding in these materials but other readers may not be. This bonding is best envisaged in the following way.

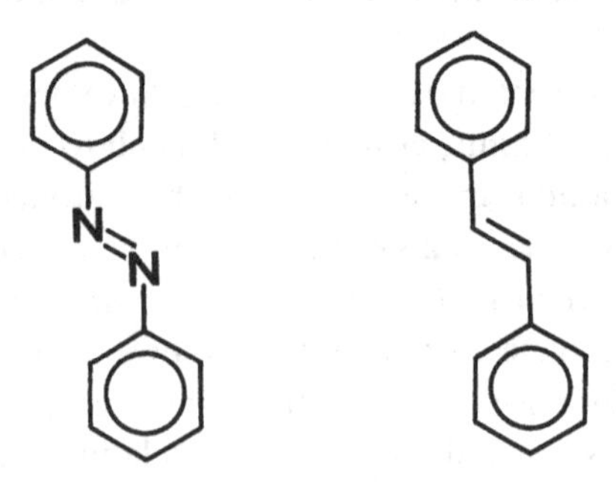

Figure 4.3. The azobenzene (left) and stilbene (right) structures.

The two nitrogen atoms or the equivalent carbon atoms in the stilbene structure are supposed to be in the trigonal hybrid state and these hybrid orbitals bond to the nearest benzene ring, to the other nitrogen atom (or carbon in the case of stilbene) and either form a lone pair in the case of azobenzene or bond to the remaining hydrogen atom in the case of stilbene. The remaining p-orbital forms a π-bond with the other nitrogen atom or equivalent carbon atom. More precisely, these orbitals become part of an extended conjugated structure involving both rings. The two rings are thus constrained to remain approximately in the same plane. This is the *trans* configuration. In the case of the azobenzene molecule it is possible to bring about a *cis* configuration in which the molecule is bent back upon itself. This latter state is slightly less stable than the *trans* state and can be produced by irradiation at about 356 nm, though the large absorption peak associated with this phenomenon is quite wide. When these groups are closely packed, as in LB films for example, the absorption peak is shifted to somewhat shorter wavelengths. In the case of the stilbene molecule the *cis* and *trans* states have equal stability.

Materials containing modifications of these groups have been used in devices intended to generate optical second harmonics and are considered in Chapter 5. In 1980 Heeseman [148] published an extensive study of amphiphilic materials containing these groups but avoided the simple straight chain structures to which reference has just been made. In 1983 Nakahara *et al.* [149] published a study of LB films made from compounds consisting of the azobenzene group with one or more long chain substituents. Blinov *et al.* [150] studied a group of compounds in which an 18 carbon chain was attached to one end of an azobenzene group and various different hydrophilic groups were attached to the other end. They succeeded in depositing these materials in both the X and Z modes, though unfortunately very little experimental information is given in their paper. Using Stark effect measurements they were able to show that, at least to some extent, the non-centrosymmetric structure remained stable. They were also able to demonstrate both piezoelectric and pyroelectric effects.

In Chapter 2, in the section devoted to X-ray diffraction, reference was made to two interesting amphiphilic azobenzene derivatives which were studied by Jones *et al.* [151] and Tredgold *et al.* [35]. The structure of these materials is shown in Figure 4.4(*a*). The shorter material gave good Y layers at a deposition pressure of 40 mN m^{-1} and the longer material also gave Y layers under these conditions, though an apparent partial

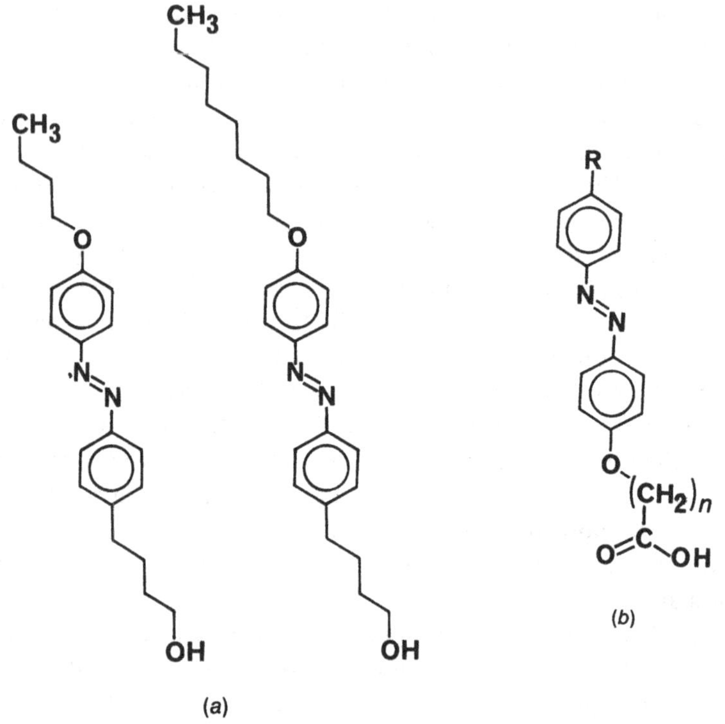

Figure 4.4. Azobenzene derivatives. (*a*) Materials discussed in [35, 151] and in the text. (*b*) Azobenzene materials discussed in [152, 153] and in the text. R is a simple hydrocarbon chain which was varied in length in the studies referred to, as was the number of CH_2 groups which is denoted by n.

Z deposition took place. This appears to be a good example of reversal after deposition, as referred to above. The remarkable characteristic of these materials is that, immediately after deposition, they appear to be homeotropic under the polarising microscope. However, on heating they undergo a phase change (at 38 °C for the shorter material and at 65 °C for the longer material) and change to a mosaic structure of birefringent domains, each being typically about 0.2 mm across. In the case of the short chain material, this latter structure remains stable on cooling to room temperature but, in the case of the longer material, cooling eventually leads to a reversion to the homeotropic phase. X-ray diffraction studies discussed in Section 2.4 suggest that the initial homeotropic phase and the birefringent phase of the short chain material may be smectic liquid crystals. These assertions are also probably true for the longer material. In Chapter 7 it is shown that films of these materials deposited by evaporation *in vacuo* behave in a similar way. In either case the birefringent phase appears to correspond to the smectic-C phase of bulk liquid crystals. The curious feature of these materials is that the low temperature phase appears to correspond to the smectic-A structure whereas the high temperature phase corresponds to the smectic-C structure; a situation which is the converse of the behaviour of ordinary bulk liquid crystals. The shorter material experiences a reduction in the layer spacing when the change to the birefringent phase takes place though this is not true in the case of the longer material. For both materials, when in the homeotropic phase, the thickness of one monolayer as measured by X-ray diffraction is less than the length of the molecule. If it is supposed that this fact is to be accounted for by molecular tilt then the tilt angle in both cases must be about 30°. It must thus be supposed that the molecules undergo some form of wobbling motion when in this phase.

Kawai *et al.* [152, 153] have studied a group of azobenzene derivatives consisting of a hydrocarbon chain, the azobenzene group, another hydrocarbon chain and a carboxylic acid termination (Figure 4.4(*b*)). Using ultraviolet absorption spectroscopy and FTIR transmission and reflection absorption spectroscopy, they were able to deduce that, in LB films, these materials resided with the molecular axis tilting at angles in excess of 20° to the substrate normal. The specific angle depends on the number of layers deposited and the lengths of the two hydrocarbon chains in each molecule. A tilt exists even for one monolayer.

If one makes use of the rather limited information available and given above one may infer that a tilt of between 20° and 30° is normal for straight chain azobenzene derivatives when deposited as LB films, even when a homeotropic phase exists. Such a structure can only be produced in a rather loosely packed film. At the moment it is an open question whether monolayers of these materials exist in the hexatic phase as is the case for fatty acids or whether the structure more nearly corresponds to the smectic-A phase. In the case of the birefringent phase described by Jones *et al.* [151] it was shown that, once this phase was established, further layers deposited by the LB technique go down in an epitaxial manner.

Many other papers dealing with azobenzene derivatives deposited by the LB technique have been published during the past few years, though, for the most part, they have only an indirect relevance to the main theme of this book. A brief selection of these papers is given here. Tanaka *et al.*

[154] and Yabe *et al.* [155] have formed LB films of amphiphilic derivatives of β-cyclodextrin and have incorporated azobenzene derivatives into them in such a way as to form a host–guest complex. Many workers have used the *cis* to *trans* change of structure referred to above and brought about by ultraviolet irradiation to change some measurable physical parameter of LB films formed from azobenzene derivatives [156–62].

4.5 Porphyrins and phthalocyanines

The basic porphyrin and phthalocyanine structures are shown in Figures 4.5(*a*) and (*b*). It will be seen that they have a strong resemblance to one another. In the case of the porphyrins, the positions bearing numbers are capable of having various groups attached to them, though many of these positions are usually occupied by hydrogen. In the case of phthalocyanines, it is obviously impossible to attach groups at the points equivalent to 5, 10, 15 or 20 in porphyrins, but groups can be attached to the peripheries of the benzene rings. These molecules both have planar structures and have a number of relatively low lying excited states. It is largely this latter property which accounts for the interest shown in LB films made from derivatives of these materials, as it seems likely that interesting devices could be made from films in which they were incorporated. These structures are, moreover, extremely stable. Porphyrins arising from natural products can survive the fractionating process applied to petroleum and the phthalocyanine group is stable to nearly 400°C. Both materials can be complexed with divalent metals, which reside at the centre of the ring. In some circumstances it is also possible to complex them with two monovalent metals or a trivalent cation which is also attached to some other ligand. A general review of the physics and chemistry of porphyrins to 1975 has been given by Falk (revised by Smith) [163].

A modification of the porphyrin group complexed with magnesium is the essential component of chlorophyll. Iron porphyrins occur in the important biomolecules haemoglobin, myoglobin, cytochrome oxidase and cytochrome c as well as a number of less commonly occurring materials. The phthalocyanine molecule was first synthesised by accident in 1928 and its structure determined in 1934. Because of their fourfold symmetry, straightforward synthetic pathways lead to either porphyrin derivatives or phthalocyanine derivatives also having fourfold symmetry. By using mixtures of starting materials in which one component bears hydrophilic side groups and is present in small quantities and a hydrophobic bearing component is present in large quantities, it is possible to produce phthalocyanines which do not have fourfold symmetry or even twofold symmetry. Chromatographic techniques can then be used to produce a relatively pure amphiphilic product. In the case of the porphyrins, it is possible to start from the unsymmetric mesoporphyrin IX dimethyl ester which is derived from haemoglobin extracted from blood. Various amphiphilic porphyrins can be made in this way. LB films of porphyrin and phthalocyanine derivatives can thus be made in three different ways.

1. Molecules which have been rendered amphiphilic by one of the techniques discussed above are deposited in the Y mode so that the planes of the molecules are nearly vertical with respect to the film plane.
2. In some cases it is possible to use materials which have fourfold symmetry. It is not entirely clear how such materials can be deposited by the LB technique.
3. The ring structure has long hydrocarbon chains attached at the corners so that they stand up on one side. These chains provide the hydrophobic component and the polarisable ring structure provides the hydrophilic moiety.

All these methods have been used successfully and examples will now be discussed. Films in which amphiphilic porphyrins alternate with some other molecule are discussed in the next chapter.

4.5.1 Amphiphilic materials

The first study of amphiphilic porphyrins at the air/water interface was made by Alexander [164] in 1937 and a further extensive study was made by Bergeron *et al.* [165] in 1967. In the mid-1980s Jones *et al.* [166–8] showed that it is possible to obtain good Y layers of protoporphyrin IX dimethyl ester and mesoporphyrin IX dimethyl ester (Figure 4.5(*c*)) both in the unmetallised form and when the material is complexed with a number of divalent metals. It was also shown that the diols derived from these materials would form good Y layers. As the alcohol group is slightly more hydrophilic than the ester group, LB films formed from the diol are rather easier to make and are more stable than those formed from the diester. These latter materials also show relatively high in-plane electrical conductivity. What is surprising is that both these groups of materials give excellent Y-type deposition but exhibit no layer structure when examined by X-ray diffraction. This is true even for materials complexed with such a relatively heavy metal as silver. Luk *et al.* [169] studied LB films of mesoporphyrin IX dimethylester indium chloride by means of electron diffraction and concluded that the film consists of crystallites formed from tilted molecules. On the basis of the evidence available, it seems likely that all these materials rearrange after deposition to form many small crystallites which do not have crystal planes corresponding to the original LB planar structure.

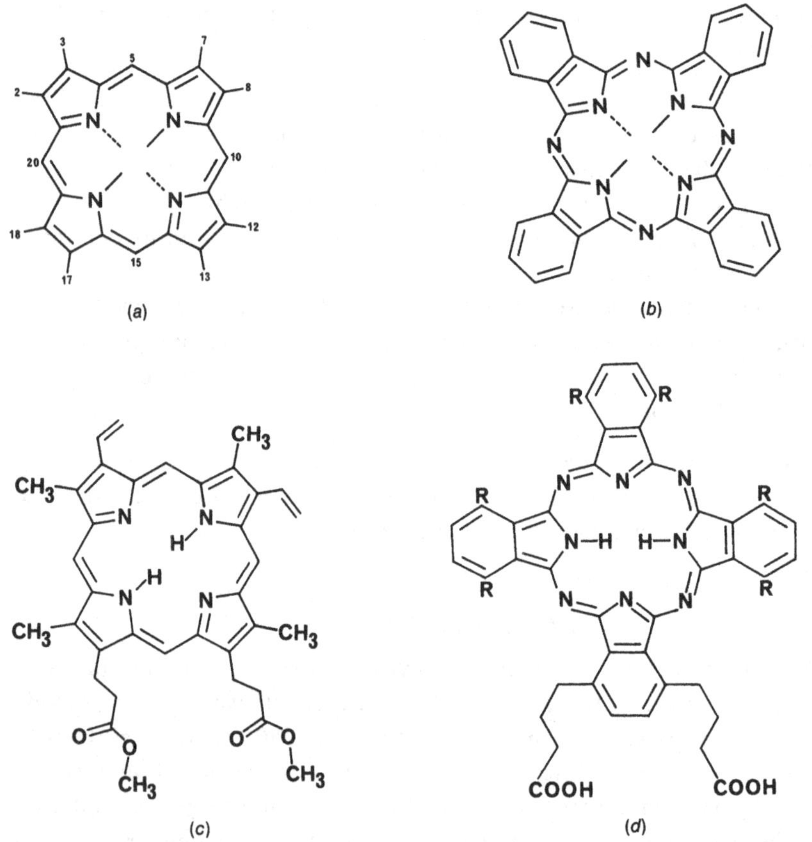

Figure 4.5 (*a*) The porphyrin ring structure. Hydrogen atoms or substituents such as $-CH_3$, $-CH{=}CH_2$ and so on are attached at the numbered points. In an earlier notation, points 5, 10, 15, 20 were denoted by α, β, γ, δ. In the parent substance, porphine, the centre of the ring is occupied by two hydrogen atoms. In the porphyrins, the centre of the ring is occupied by a metal atom, which is usually divalent. (*b*) The phthalocyanine ring. As in the porphyrin ring, a divalent metal can be accommodated at the centre of this structure. (*c*) Protoporhyrin IX dimethyl ester. (*d*) An amphiphilic phthalocyanine as discussed in [173] and the text. R can be $-CH_8H_{17}$, $-C_9H_{19}$ or $-C_{10}H_{21}$.

A variety of different methods of synthesising non-symmetric porphyrins are given in [163] but none of these are easy and, in consequence, there are very few papers in the literature concerning synthetic amphiphilic porphyrins. An amphiphilic cobalt tetrapyridylporphyrin derivative was produced by van Galen and Majda [170]. Here the pyridyl groups are attached to points 5, 10, 15 and 20 with the nitrogens at the points remote from the porphyrin ring. One of the pyridyl groups is quaternerised and a 16 carbon chain is attached to this point. They succeeded in depositing this material on a gold coated substrate by the self-assembly technique. They do not appear to have attempted to use this material in the LB context. Nagamura *et al.* [171] made a similar compound except that the three groups not bearing a long hydrocarbon chain were simple phenyl groups rather than pyridyl groups. They were able to deposit six layers in the Y mode but have given no data about structure or layer quality.

Baker *et al.* [172] were the first to attempt to produce an amphiphilic phthalocyanine but their material would only deposit on the up stroke. Cook *et al.* [173–5] produced a truly amphiphilic series of phthalocyanines, the most interesting of which is illustrated in Figure 4.5(*d*). This material was shown by McKeown *et al.* [176] using X-ray diffraction to form reasonably good Y layers. Three orders of Bragg peaks were observed, which is totally different from the behaviour of the amphiphilic porphyrins discussed above but indicates a poor degree of ordering as compared with fatty acids. Immediately after deposition the bilayer spacing is 4.1 nm but, on heating to 127.5°C, there is an irreversible phase change which reduces the bilayer spacing to 3.6 nm and brings about a marked change in the absorption spectra in the visible region. Cook *et al.* [175] have studied the properties of the phthalocyanine which is the non-amphiphilic analogue of this material and deduce that, in the initial phase, the molecular planes are nearly vertical and stack in a herring bone pattern. (It is supposed that the herring bone pattern is viewed as being projected on to a plane vertical with respect to the plane of the substrate.) When the material has been heated above 127.5°C it takes on a columnar discotic structure (see Chapter 7) with the axis of the columns in the plane of the film. As the molecules do not have sixfold symmetry, it is difficult to understand exactly what happens in this case but, presumably, the structure must be more complex than one simply consisting of hexagonally packed columns.

4.5.2 Symmetric molecules

Turning now to non-amphiphilic porphyrins and phthalocyanines, one finds a considerable number of papers dealing with LB films made from the latter class of material though it is very difficult to form LB films from symmetric porphyrins, as was shown by Bull and Bulkowski [177]. Un-substituted phthalocyanines are not soluble except in concentrated sulphuric acid and thus initial studies of substitution were largely made with a view to producing materials which would dissolve in user friendly solvents. The first success in forming LB films of phthalocyanines was obtained by Baker *et al.* [178] who worked with tetra-t-butylphthalocyanine (Figure 4.6) in its metal free form. Though it was possible to form films at the air/water interface and deposit films on solid substrates, the deposition ratio was indeterminate and the structure of the films was not characterised. More satisfactory results were obtained by the same group [179] using a similar material complexed with silicon dichloride. Studies of dichroism indicated that the planes of molecules were probably vertical with respect to the substrate and tended to be oriented along the dipping direction.

If the molecules stand on edge and are oriented by the dipping process, there will exist an optical anisotropy which will manifest itself in dichroism or, if the material is thick enough, in birefringence. On the other hand, if the molecules lie flat, no optical anisotropy will be shown when the material is examined by transmitted light. Thus polarising

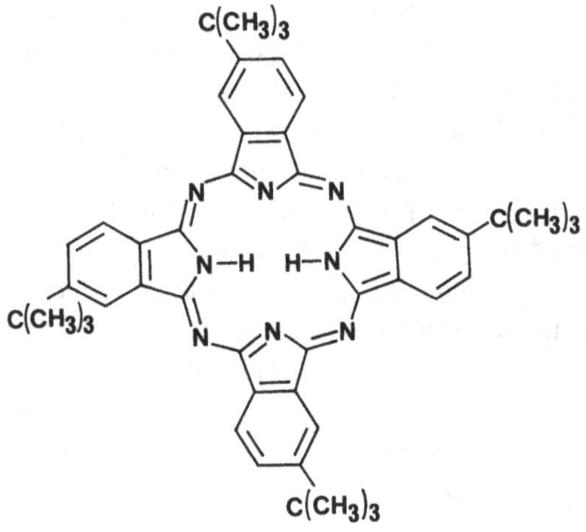

Figure 4.6. Structure of tetra-t-butylphthalocyanine as discussed in [178].

microscopy and low angle X-ray diffraction are the two tools which have been most used to study this class of material. Unfortunately, very little work has been published which makes use of both these techniques. As will be seen from the work summarised below, true LB deposition usually leads to an edge-on structure, whereas related techniques can lead to a structure in which the molecular planes lie parallel to the substrate. Thus, for example, Fujiki *et al.* [180] studied cast films of materials bearing four octadecyl chains attached to the ring by amide bonds and found that the ring structures were parallel to the substrate. Nakahara *et al.* [181] studied systems in which eight alkyl chains were attached directly to the ring and formed their films by the horizontal lifting method in which the substrate is brought into contact with the water surface in a nearly horizontal posture. They also found that the molecules were parallel to the substrate.

On the other hand, Ogawa *et al.* [182] studied the copper complex of a material bearing four butyl chains attached to the ring by ester bonds. They obtained good Y layers by normal dipping procedures and obtained a high degree of dichroism, indicating that the planes of the molecules lie in the dipping direction and are normal to the substrate. Nichogi *et al.* [183] studied tetrapentoxyphthalocyanine complexed with lead and obtained a structure in which the molecular planes are normal to the substrate. Fryer *et al.* [184] used high resolution electron microscopy and electron diffraction to study copper tetra-t-butylphthalocyanine. They found that LB films of this material consisted of islands of crystalline material having the molecular planes normal to the substrate which were embedded in disordered material. The same material was studied in detail by Brynda *et al.* [185, 186] who showed that the molecular planes lie at 14° to the normal to the plane, that the individual planes consist of small crystallites not more than 20 nm in extent and that, surprisingly, there is no epitaxy between one layer and another. Thus, on the basis of the existing evidence, true LB films of simple symmetrically substituted phthalocyanines form structures in which the molecular plane is normal or nearly normal to the substrate, whereas cast films or films formed by the horizontal dipping method can have the molecular planes parallel to the substrate. Phthalocyanines having more bulky and complex substituents have been studied for example by Fukui *et al.* [187] and by Pace *et al.* [188] but it is difficult to arrive at any useful generalisations about such materials.

Several studies have been made of LB films formed from dimers of phthalocyanines. Shutt *et al.* [189, 190] formed a dimer in which a germanium atom resides at the centre of one ring and a silicon atom at the other. These two atoms are linked by an oxygen atom and their requirements for tetravalence are satisfied by an OH group, attached to the germanium atom, and a six carbon alkoxy group, attached to the silicon atom. Liu *et al.* [191] formed LB films from lutetium diphthalocyanine. In both these cases the planes of the molecules were parallel to the substrate. The principle of linking phthalocyanines in this way has been exploited to form polymers which can be dissolved in organic solvents and spread at the air/water interface. These materials will be discussed in Chapter 5.

4.5.3 Rings with four long chains attached

The synthesis of porphyrins which bear four long hydrophobic chains and whose hydrophilic moieties are associated with the ring structure has been pursued by Ruaudel-Teixier *et al.* [192] and Lesieur *et al.* [193]. These materials are based on the tetraphenyl porphyrin structure in which phenyl groups are attached at points 5, 10, 15 and 20. Two types of molecule were produced. In one the phenyl rings were replaced by pyridine rings quaternised with $C_{20}H_{41}Br$. In the other the phenyl rings were substituted in the *para* position with an α branched docosanoic acid. These otherwise interesting materials have not, however, led to the formation of ordered multilayers and will thus not be discussed at length here. Palacin *et al.* [194, 195] applied the same general principle to phthalocyanines and obtained stable films at the air/water interface which could produce multilayers by the LB technique.

4.6 Envoi

In this chapter an attempt has been made to discuss ordered structures made using the LB technique and employing relatively simple molecules all of one kind. In the next chapter, films made from preformed polymers, from polymerisable small molecules and from alternating layers of two distinct kinds of molecule will be discussed. Once again the emphasis will be on structure and the characterisation of order. The discussion of lipids and lipid-like materials is deferred until Chapter 8 which will discuss biomembranes and bioactive molecules. At this stage the reader may well ask 'What of the vast number of other simple materials from which LB films have been made?'

It is an unfortunate fact that, whereas many hundreds of materials have been used to form LB films, in the majority of cases no serious effort has been to characterise the film structure or even to show that a regular layer structure has been achieved. The example of amphiphilic porphyrins discussed above is sufficient to emphasise the need for proper characterisation. It will be recalled that mesoporphyrin IX dimethyl ester and the equivalent diol can, when complexed with a large variety of divalent metals, form excellent apparent Y layers with a deposition ratio very near unity on both up and down strokes. However, it is quite impossible to demonstrate the existence of a regular layer structure using X-ray diffraction techniques. Analogous situations involving two kinds of molecule will be discussed in Chapter 5.

In the ideal situation any supposed LB layer structure would be examined by low angle X-ray diffraction, electron diffraction and by the polarising optical microscope. It is unfortunate that this procedure is rarely carried out. Thus in this chapter only those cases have been discussed where at least some effort has been made to determine the structure and degree of order of the film.

A recent book by Ulman [197] gives a reasonably comprehensive account of LB work up to December 1989 and references many papers not discussed here.

5

More complex structures formed by the Langmuir–Blodgett technique

5.1 Introduction

In this chapter we turn to the study of LB films formed from polymers and LB films consisting of alternate layers of two different amphiphiles. In principle, of course, it would be possible to superimpose successive layers of three or more distinct amphiphiles but little has been done in this direction. However, a few examples of more complex alternating structures will be considered.

Polymer LB films naturally divide into the following categories.

1. Systems in which a multilayer structure is formed from molecules containing one or more double bonds and in which polymerisation is subsequently initiated by irradiation by γ-rays, ultraviolet light or an electron beam.
2. Systems similar to the above but in which the constituent monomers contain the diacetylene group.
3. Multilayers formed from polymers bearing both hydrophilic and hydrophobic side groups which are spread as *polymers* at the air/water interface and are subsequently deposited on a substrate by the LB technique.
4. Rigid rod polymers which have both hydrophilic and hydrophobic characteristics and which are capable of residing with the rod axis horizontal at the air/water interface and which can be deposited on a solid substrate by the LB technique.

5.2 Post-formed polymers made from monomers containing one or more double bonds

Studies of polymerisation at the air/water interface have been made repeatedly over the years and an account of early work in this field is given by Gaines [14]. Here the discussion will be largely confined to polymerisation carried out after deposition. A study of this phenomenon has been of abiding interest to polymer chemists as it is possible to arrange the monomers so that the polymerisable groups are adjacent to one another and to monitor changes in film structure arising from polymerisation. The system under study is thus far more under the control of the investigator than is normally the case. If the double bond is to respond readily to activation by ultraviolet light, it is necessary for such a bond to be close to a part of the molecule such as a carboxylic group, an ester group or a benzene ring so as to form a conjugated structure. However, as will be seen below, isolated double bonds can to some extent respond to irradiation by ultraviolet light and lead to polymerisation, particularly if they link the end carbon and penultimate carbon atom in a chain. The first group of results presented here all involve double bonds near the hydrophilic end of the molecule and thus fall into the category in which the double bond is not isolated.

An account of the first work making use of this idea was published by Cemel *et al.* [198] in 1972 using vinyl stearate (Figure 5.1(*a*)). They formed multilayers of this material and then initiated polymerisation by irradiation with γ-rays from ^{60}Co. They found that the presence of oxygen inhibited polymerisation and thus carried out this process in a nitrogen atmosphere. Best results were obtained when the irradiation was carried out at either 0°C or −78°C and the material was then allowed to polymerise at 20°C. The process of polymerisation was monitored by observing the strength of the infrared absorption band at 948 cm^{-1} associated with vinyl hydrogen deformation. Under the appropriate conditions, a very high level of polymerisation could be achieved. Clearly, however, such a process could not lead to complete polymerisation as individual monomers must become isolated here and there. The authors claim that, irrespective of the exact dipping conditions, the multilayers arrange themselves in an X conformation with the monolayer not the bilayer as the basic repeat unit. This claim is based on the repeat distance as measured by X-ray diffraction, which corresponds approximately to the length of a monomer. In the two decades since the publication of this paper it has become apparent that simple straight chain compounds normally deposit in the Y mode and it thus seems more likely that this material too deposits in this way, but with a large tilt angle possibly combined with a degree of interdigitation.

In 1977 Naegele *et al.* [199] investigated the polymerisation of cadmium octadecyl fumarate (Figure 5.1(*b*)). In the case of this material

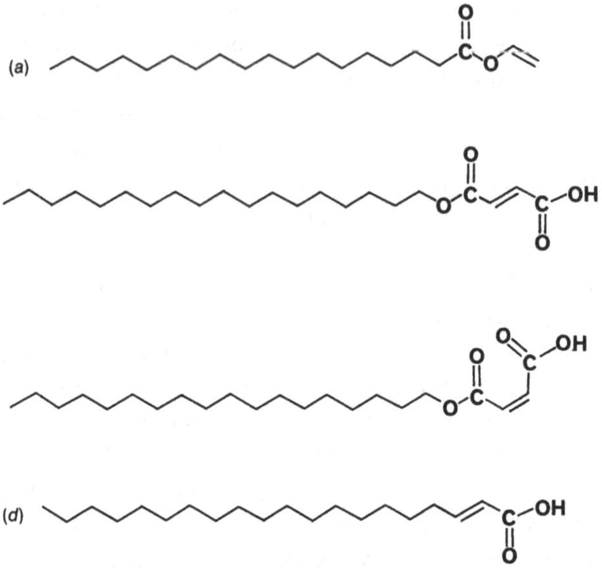

Figure 5.1. Structures of (*a*) vinyl stearate, (*b*) octadecyl fumarate, (*c*) octadecyl maleate and (*d*) 2-eicosenoic acid.

it is possible to initiate polymerisation by irradiation by ultraviolet light. Good Y layers were obtained and studied by infrared spectroscopy, X-ray diffraction and transmission electron diffraction. For multilayers prior to polymerisation, it was possible to show that, in the case of films up to about ten layers thick, the structure was hexagonal, but for thicker layers the structure became tilted with monoclinic symmetry and a tilt angle of about 37°. The repeat distance in the direction normal to the plane of the film is 5.8 nm in the case of the hexagonal structure and 4.3 nm in the case of the monoclinic structure. The repeat distance for fully polymerised film is 5.5 nm, which indicates that, in this case, the molecule axes are nearly normal to the film plane. If the layer spacing is monitored during the polymerisation process, it is found that there is a continuous change in lattice spacing for the hexagonal structure but that, for the monoclinic structure, there is a discontinuous transition at about 40% conversion. It should be pointed out that the presence of the cadmium ions, which have a large X-ray scattering factor, in the hydrophilic/hydrophilic region of the film make it possible to obtain good X-ray diffraction patterns for films containing only a few monolayers. Rabe *et al.* [200] studied the polymerisation of both cadmium octadecyl fumarate and its isomer cadmium octadecyl maleate (Figure 5.1(*c*)), making use of infrared spectroscopy. Prior to polymerisation, they claim that the ODF films exist with the molecular chains almost normal to the substrate (in contrast to the results of Naegele *et al.* [199]) whereas the ODM molecules have a considerable tilt. After polymerisation using ultraviolet light it appears that, irrespective of the starting material, the polymers produced are identical. These workers found that it was necessary to use a filter to cut out wavelengths shorter than 254 nm to avoid destruction of the film by shorter wavelength ultraviolet and this is, indeed, a precaution which needs to be taken in all analogous experiments.

Laschewsky *et al.* [201] carried out an important series of exploratory experiments in which a wide variety of possible amphiphilic monomers were deposited by the LB technique and polymerised by irradiation by ultraviolet light. With one exception, all these materials were spread on a pure water subphase which did not contain a divalent cation. The degree of polymerisation as a function of time was monitored by the change in ultraviolet absorption spectra. Layer spacings and their change with polymerisation were measured using X-ray diffraction. It is impossible to give here a complete summary of the results reported in this paper. Of particular interest, however, are the results obtained from the monoesters of three different perfluorinated alcohols. All three formed good multilayers and polymerised readily under the influence of ultraviolet light. There is some evidence that, for some of the materials discussed above, the equilibrium spacing of the polymer sub-units is slightly larger than the corresponding spacing for the monomers in a multilayer. Thus the slightly larger cross section of the perfluorinated tail as compared with the normal hydrocarbon tail might facilitate polymerisation. Laschewsky *et al.* [202] studied octadecyl fumarate and 2-eicosenoic acid (Figure 5.1(*d*)) deposited from a subphase of pure water. Low angle X-ray diffraction results obtained from multilayers of the latter material indicate that the structure of these layers is less ordered than those of other fatty acids.

Laschewsky and Ringsdorf [203] studied the deposition and polymerisation of multilayers of alcohols and acids incorporating the diene group, $-CH{=}CH-CH{=}CH-$ at the hydrophilic end of the molecule. The carbon atoms in this group are conventionally labelled from 1 to 4 starting from the hydrophilic end. Ultraviolet absorption studies of the polymerisation process indicate that initial polymerisation probably links atoms 1 and 4 in adjacent molecules respectively, thus creating a new double bond between atoms 2 and 3. This new bond can now take part in a cross linking process which leads to a totally insoluble product. However this second process involving an isolated double bond takes place more slowly than the initial polymerisation. Systematic but small changes in layer spacings are observed as polymerisation proceeds and these are attributed to changes in molecular tilt.

The materials so far discussed in this section all have the double bonds near the hydrophilic end of the molecule. In contrast to this Barraud *et al.* [204, 205] published a series of papers describing the properties of ω-tricosenoic acid which is a straight chain carboxylic acid containing 22 carbon atoms in which the penultimate and final carbon atoms at the hydrophobic end are joined by a double bond. The material does not polymerise so rapidly when irradiated by ultraviolet light as do the substances discussed above but is readily polymerised when bombarded by an electron beam. It is an exceptionally easy material to deposit in multilayers and was at one time thus thought to be an ideal material for the fabrication of electron beam resists. It has not as yet found a practical application in this context. This is due to two factors. On the one hand, it is found that the monomer to monomer distance decreases slightly on polymerisation and thus over distances involving 50 or more molecules the deformation of the film is such that cracks appear. On the other hand, spun polymer resists are capable of giving a degree of definition which more than caters for present day needs and there exist other more important limitations to the definition obtainable than those arising from the nature of the resist material. However, the phase diagram of this material is such that it deposits with great rapidity at room temperature and has thus been intensively studied by workers not actually interested in the problem of polymerisation.

Uchida *et al.* have studied the deposition and photo-polymerisation of relatively complex amphiphilic compounds having two hydrophobic chains attached to a single hydrophilic head group [206, 207]. They have synthesised and deposited several interesting materials having one double bond near the head group, one double bond in one of the hydrophobic tails or double bonds in both tails with a view to obtaining materials suitable for forming stable bio-compatible coatings for artificial organs. It is difficult to assess the degree of order in the films obtained, as the only measurements made which bear directly on structure involved X-ray diffraction from a beam impinging normally on a thick film. The most important outcome of this work is the formation of stable multilayers having a hydrophilic outer surface. This result was obtained using a material having double bonds in both tail groups.

5.3 Diacetylenes

The study of three-dimensional crystalline polymers formed from diacetylenes had already made substantial progress before attention was turned to LB films of these materials. A number of symmetrical compounds having the diacetylene group at the centre have been studied and the book edited by Bloor and Chance [208] based on the proceedings of a conference funded by NATO gives a good general picture of progress made up to 1985.

Most of the diacetylene based materials studied in the LB context have been diynoic acids, the structure of which is shown in Figure 5.2. The notation for defining these acids is given in the caption for that figure. If these materials are deposited as LB multilayers, polymerisation can be induced either by thermal or optical means. Figure 5.3 gives a schematic representation of the rearrangement of the material on polymerisation. This subject has been extensively studied by Tieke and various collaborators [209–13]. Other papers devoted to this general topic will be referred to as appropriate. However, this field has attracted so much attention that it would not be practicable to give a comprehensive list of references. It is possible to vary *m*, to dip over a subphase

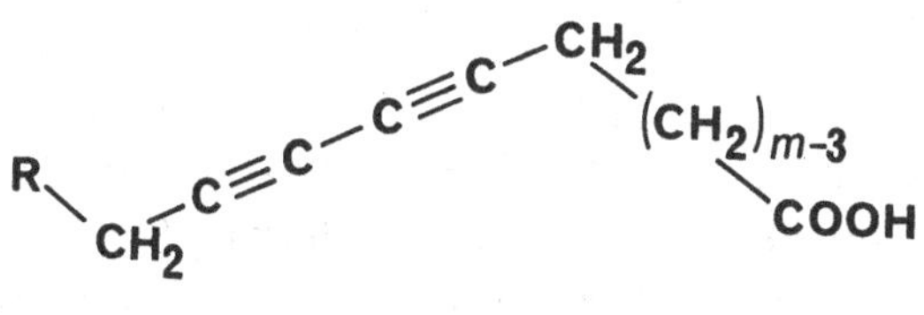

Figure 5.2. Tricosa-*m*, *n*-diynoic acid (a diacetylene fatty acid). This is a schematic diagram to illustrate the definition of *m* and *n*. $n = m + 2$.

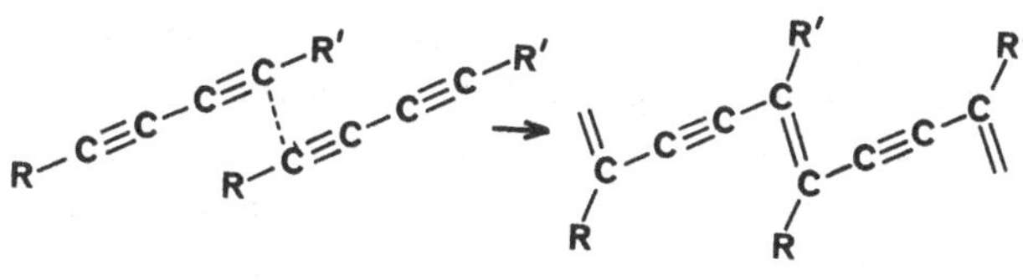

Figure 5.3. Schematic diagram to illustrate the polymerisation of diacetylenes.

either containing or not containing divalent cations, to vary the pH of the subphase and to bring about polymerisation either before or after dipping. Furthermore, the temperature at which polymerisation takes place and the wavelength of light inducing polymerisation can also be varied. It is thus not surprising that this subject has an extensive literature. In addition to structural studies, a great deal of work has also been devoted to the characterisation of the optical and electrical properties of this family of materials, though these topics, being only of marginal relevance to the main themes explored in this book, will not be discussed at any length here.

Studies made by means of electron diffraction indicate that, in LB films prior to polymerisation, the diacetylene monomers are vertical with respect to the plane of the film. After polymerisation the film has monoclinic symmetry and the side chain axes have a tilt of about 30° with respect to the normal to the plane of the film, as one might indeed expect from the behaviour illustrated in Figure 5.3. Polymerisation takes place via initial formation of a dimer and this requires a thermal activation energy of about 1 eV. Photo-polymerisation will certainly take place on irradiation by light having a wavelength of 300 nm (corresponding to 4 eV), but slow photo-polymerisation can take place under the influence of light having a much longer wavelength. As far as I am aware no systematic study of the action spectra of polymerisation has ever been carried out. Once the dimer has been formed the subsequent reaction is exothermic, with release of about 0.6 eV per monomer added to the chain. Both diradicals and dicarbenes (Figure 5.4) are thought to be formed and chains are finally terminated by rearrangement of the structure shown in the last part of Figure 5.4. As the process of polymerisation leads to a molecular tilt, it is likely to be cooperative in nature and a given region will take on a tilt in some arbitrary direction. One would thus expect a structure to form in which many different domains will be formed, having different orientations distinguishable from one another under the polarising microscope. This phenomenon is in fact observed, the size of the domains being determined by a variety of factors and extending for distances ranging from a few micrometres to a few hundred micrometres in the film plane. It is this factor that has, above all, hindered the practical application of such films. Tieke and Weiss [213] showed that the nature of the dipping solvent was important in determining the domain size in a study in which cadmium ions were present in the subphase. Most of the results reported in this paper were obtained using material in which $m = 10$ and the total number of carbon atoms

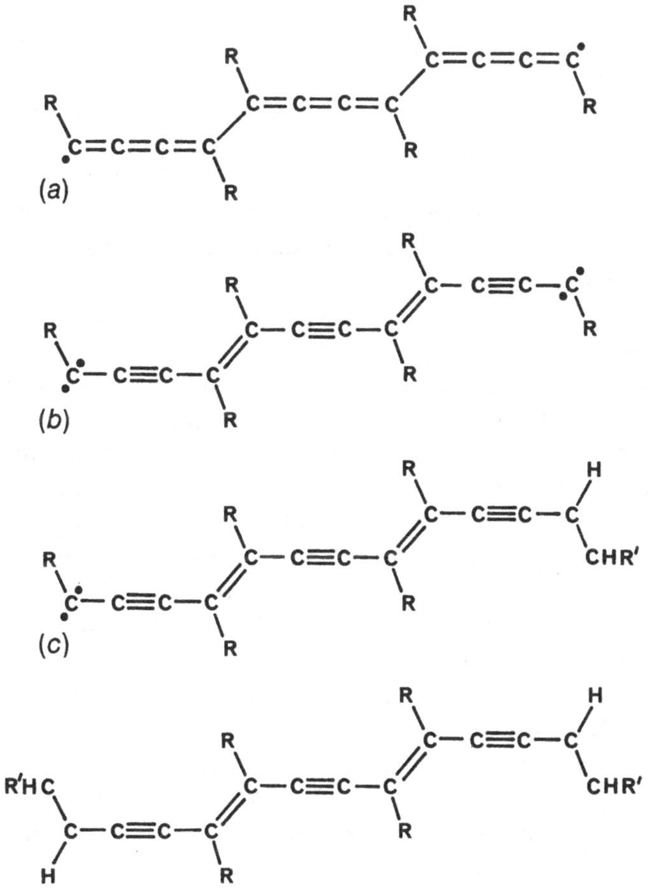

Figure 5.4. The polymerisation of diacetylenes. (*a*), (*b*) and (*c*) are various possible intermediate structures. A dot denotes a free radical and two dots a free diradical. This figure is reproduced by kind permission of the authors and publishers from *Polydiacetylenes, Synthesis, Structure and Electronic Properties* by Bloor, D. and Chance, R.R. 1985 (Kluwer Academic, Dordrecht).

in the molecule was 25. For LB films formed from cadmium salts of this material, a change in the absorption spectra takes place at 59°C though an activation process clearly is involved as this change is only observed after a considerable annealing time. This phase change has the effect of making the material change from appearing blue in colour to appearing red as the temperature is increased.

The so-called blue and red phases of polydiacetylenes have been extensively investigated. Thus, for example, Tomioka *et al.* [214] studied the behaviour of partially and totally polymerised monolayers of the diynoic acid having $m=10$, $n=12$ and found that, as the surface pressure was increased, so the colour changed from blue to red. Their extensive investigation reached the conclusion that the colour was brought about by the degree of packing, the low density form appearing blue and the high density form red. Studies of reflection spectra from partially polymerised material made by these authors, in which spectra were obtained for various surface pressures, showed that the change of spectra was continuous with pressure. These authors make a convincing case for the view that compression of the monolayer brings about a distortion of the polymer chain which reduces the delocalisation of the π electrons, which can be shown to lead to the changes of the absorption and reflection spectra which are observed. Tamura *et al.* [215] made observations on material having $m=10$ and $n=12$ which confirmed the results of the authors quoted above. They also carried out differential scanning calorimetry and concluded that, in the LB multilayers studied, the colour change from blue to red is brought about by disorder in the side chains and the resultant effect on the structure of the main chain.

A number of studies have been made of LB films of materials having two hydrophobic tails each containing a diacetylene group, but little has been done to characterise their order. Various amphiphilic materials having head groups other than a carboxylic acid have been studied. An interesting recent example is a material containing ferrocene, the properties of which were studied by Fukuda *et al.* [216]. Ahmed *et al.* [217] have synthesised and studied acids containing two distinct diacetylene groups separated by carbon chains of various lengths. These materials exhibit curious isotherms at the air/water interface, in which a partial apparent collapse at intermediate pressures precedes a normal build up of pressure with decreasing area and final collapse at high pressures. This behaviour is not yet understood.

The conjugated polymeric structure of the diacetylenes produces a planar structure which exhibits a relatively high electronic conductivity and a high third order hyperpolarisability, χ_3, and thus they have considerable potential in technical applications. These phenomena, however, fall outside the scope of this book. Furthermore, the polycrystalline nature of films discussed above has so far prevented realisation of these potentialities.

5.4 Preformed polymers

It will have become apparent to the reader that LB films formed from simple rod-like molecules, or from such molecules polymerised after deposition, tend to consist of well characterised layers which, however, have a lot of unwanted structure in the film plane. This property is unfortunate, as many possible applications of such films require a uniform structure within the plane. In an ideal system it would be possible to form LB films which consisted of true three-dimensional crystals, but so far this has proved impossible. Another possibility is to produce a system in which the structure within the plane is so disordered that there exist no structural features large enough to cause problems in a given proposed application. Thus, for example, in three-dimensional materials both inorganic glasses and many polymers are capable of transmitting light without appreciable scattering for substantial distances. The extreme example of this property is the propagation of light for distances measured in kilometres by optical fibres. So far no one has succeeded in producing an LB multilayer in which the structure within one layer is a true glass. However, another possibility presents itself. Suppose one were to form LB films from an existing polymer which has hydrophilic side groups on one side of the main chain and hydrophobic groups on the other. There then seems a possibility that the material will form a good layer structure but that one will achieve a fairly random arrangement within each layer. The study of monolayers of such materials at the air/water interface has a considerable history and is discussed in the book by Gaines [14]. However, as far as I am aware, the first successful attempt to form real multilayers of such materials was reported in 1982 by Tredgold and Winter [218]. They studied copolymers of maleic anhydride and various vinyl derivatives. In this initial study the maleic anhydride gradually hydrolysed at the air/water interface but in subsequent work the maleic anhydride rings were opened by reacting the polymer with an alcohol, in the simplest case methanol, but also by a wide range of longer chain and more complex alcohols. The maleic anhydride/vinyl copolymer system is known to give an exceptionally regular structure and it was thus possible to produce copolymers of the type illustrated in Figure 5.5 in which the group R^2 is introduced as an alcohol and the group R^1 is introduced as a vinyl derivative.

This initial work was followed by an extensive study of LB films formed in the general manner explained above [219–32]. Particular emphasis was placed on materials in which either R^1 or R^2 were

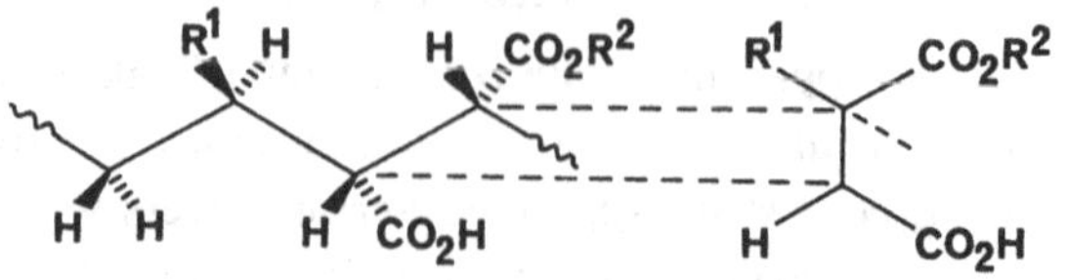

Figure 5.5. Maleic anhydride vinyl derivative copolymers. The left hand part of the diagram illustrates a section of the polymer chain viewed from the side. The right hand part of the diagram illustrates the chain viewed end on from the right. ▼ denotes a bond protruding from the plane of the diagram and ⦙ denotes a bond receding from the plane of the diagram. R^2 is a group introduced as an alcohol and R^1 is a group introduced as a vinyl derivative.

azobenzene derivatives, hemicyanines or materials likely to form a mesophase. It would not be appropriate to give an extensive account of these studies here, but certain generalisations are worth recording.

1. It was possible to form stable layers at the air/water interface and hence form LB multilayers when both R^1 and R^2 consisted of quite short chains. An example of such a material is styrene/maleic anhydride copolymer hydrolysed so as to open up the maleic anhydride rings.

2. Materials in which R^2 was either long or bulky packed poorly at the air/water interface and it was sometimes impossible to form multilayers from such materials. The reason for this behaviour is apparent if reference is made to Figure 5.5. The hydrophilic carboxylic acid group is attracted to the water surface and hence the group R^2 is forced downward towards the water. If this group is only partly hydrophobic, as is the case for a methyl ester, there is no difficulty about such an arrangement taking place. However, if this latter group is large and hydrophobic, it will cause a change in the conformation of the chain so that both R^1 and R^2 point upwards. In fact it seems likely that a mixture of these conformations can exist in one chain, causing a disruption to regular packing.

3. Materials in which R^2 is small (a methyl group for example) and R^1 is a long simple hydrocarbon chain pack well and form well ordered structures. Tredgold *et al.* [222] showed that, for such materials, it was possible to obtain three clearly defined Bragg peaks when a multilayer was studied by X-ray diffraction (see Figure 5.6). On the other hand, materials having a more complex R^1 group or having a short R^1 or a long R^2 rarely exhibited more than one Bragg peak. The analysis given in Section 2.2 is relevant to these results. The fact that we are dealing

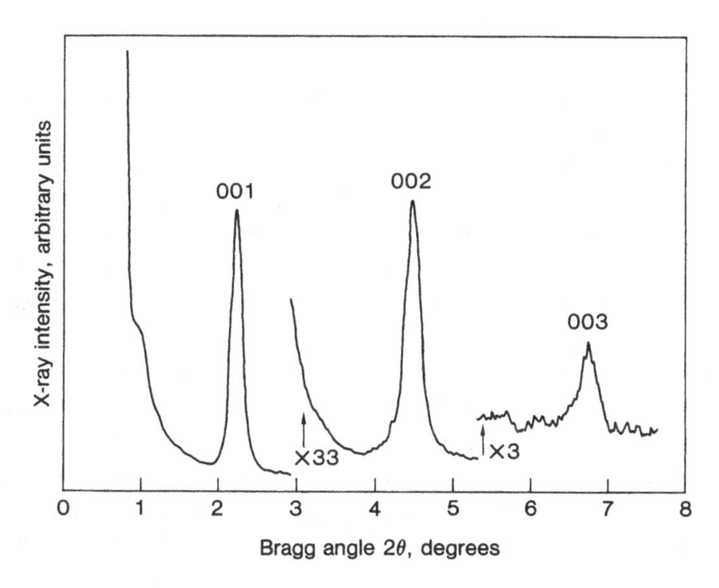

Figure 5.6. X-ray diffraction trace obtained from a multilayer of poly (octadec-1-ene-maleic anhydride). (Reproduced from Tredgold, R.H., Vickers, A.J., Hoorfar, A., Hodge, P. and Khoshdel E. 1985 *J. Phys. D: Appl. Phys.* **18** 1139–45 by kind permission of the Institute of Physics).

with a polydispersive material leads to a structure in which polymer chains of widely different lengths are compressed together and then deposited in successive layers. Voids are thus certain to exist. Furthermore it is likely that, here and there, polymer chains will cross over one another. Both these processes will produce defects in the regular layer structure. Reference to Section 2.2 shows that, for a layer structure having long range order, successive Bragg peaks will decrease rapidly in amplitude but remain constant in width, as shown in Equation (2.14). However, where true long range order no longer exists, successive Bragg peaks decrease in magnitude and increase in width as shown in Equation (2.25). The results shown in Figure 5.6 tend more to the latter behaviour than to the former. An absence of long range order is even more probable in the case of those materials where only one Bragg peak appears.

Studies of the waveguiding of light, in multilayers of some polymers of the general category just discussed, made by Tredgold *et al.* [227] showed that it was possible to propagate light of wavelength 633 nm in a film 0.5 μm thick with an attenuation of 10 dB cm^{-1}. This attenuation is, of course, large compared with that of many materials but is very small compared with other LB materials. Thus the aim of producing a layer structure with a random polymer arrangement within each layer has been, at least in part, achieved. Other applications of this family of polymers are discussed in Section 5.6.

Another approach to the fabrication of LB films from preformed polymers was taken by Ringsdorf and his collaborators [233–7]. Here a hydrophobic main chain is formed by reacting monomers terminated by a vinyl group. This group is connected to the main side group via an amphiphilic spacer group. Thus the main chain is probably immersed in the water, the spacer groups are located at the air/water interface and the side group is above the water (Figure 5.7). This type of structure clearly derives from the type of side chain polymer which is employed to obtain mesogenic properties, where a long flexible spacer is required in order to allow the mesogenic side groups to interact. The side groups studied included lipid-like structures, typical mesogenic groups and perfluorinated hydrocarbon chains. These latter materials are of interest

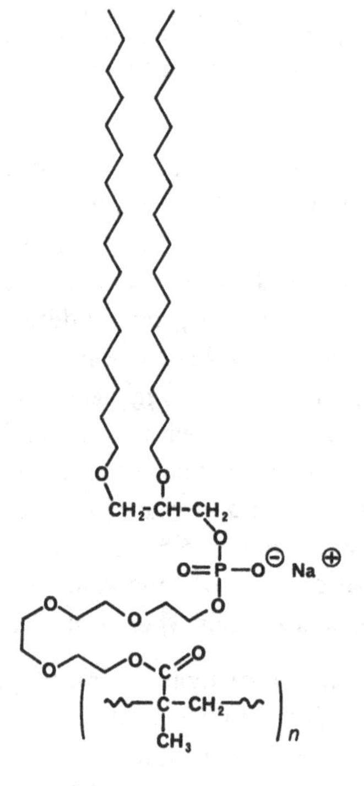

Figure 5.7. A polymer in which the main chain is attached to the hydrophobic side chains by a hydrophilic moiety.

as the perfluorinated groups were shown by infrared studies to tilt with respect to the normal to the plane of the film whereas the analogous ordinary hydrocarbon chains did not [236]. This behaviour is similar to that of perfluorinated carboxylic acids deposited by evaporation *in vacuo* [145]. The degree of order obtained by this technique is not high, as only one Bragg peak is obtained in low angle X-ray diffraction studies.

The polymethacrylates have been studied as LB materials by several groups (Figure 5.8). Gabrielli *et al.* ([238] and a number of subsequent papers) studied the behaviour of polymethylmethacrylate at the air/water interface and Mumby *et al.* [239] studied polyoctadecylmethacrylate formed into LB multilayers. These latter authors claim to have achieved Z-type deposition but, in view of the rarity of true Z-type deposition and the fact that they did not use X-ray diffraction to determine layer spacing, this claim must be viewed with caution. Naito [240] formed LB multilayers of polyisobutylmethacrylate but failed to obtain any Bragg peaks in the low angle diffraction mode. This result suggests that a largely disordered film was produced which, in view of the structure of this material, is hardly surprising. Brinkhuis and Schouten [241] studied the behaviour of largely isotactic polymethylmethacrylate at the

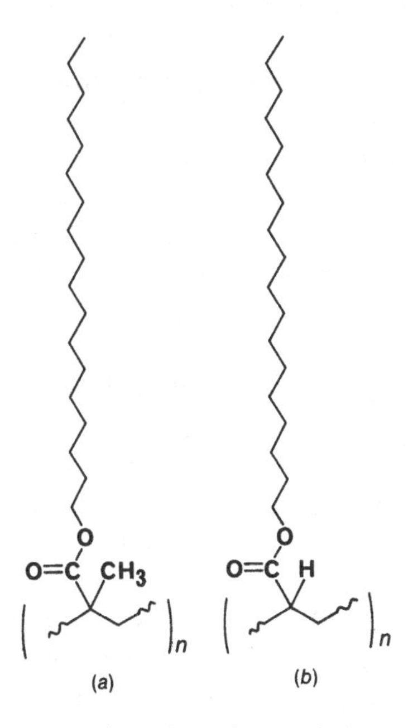

Figure 5.8. (*a*) A polymethacrylate and (*b*) a polyacrylate.

air/water interface and also when this material was formed into LB films. (An isotactic polymer is one in which the 'handedness' of each unit forming the polymer is the same as that of the others.) These authors deduced that, at a surface pressure of 20 mN m^{-1}, there is a phase change in which the material takes on a double helix formation. Very strongly oriented LB films could be formed from this material with, it is to be presumed, the helices lying parallel to one another.

Several studies have been made of LB films of esters of naturally occurring polysaccharides. Kawaguchi *et al.* [242] formed long chain esters of cellulose which, however, could only be formed into multilayers by the horizontal lifting technique. Schoondorp *et al.* [243] studied LB multilayers of esters of amylose and showed that materials with short alkyl side chains have a helical conformation at the air/water interface and that this structure can be transferred into multilayers. As in the case of the isotactic polymethylmethacrylate, the helical structure appears to lead to an oriented structure in the LB film. These two families of materials are illustrated in Figure 5.9.

In the last few years many papers have appeared describing LB films formed from preformed polymers and it is not really practicable to list them all here. A brief selection of those in which an attempt has been made to characterise the degree of order is given below. Oguchi and his collaborators [244–6] studied acetalised polyvinyl alcohols and showed by means of ellipsometry that the film thickness increases linearly with the number of layers deposited. Lupo *et al.* [247] studied polyamides bearing long side chains. Using infrared spectroscopy as their principal diagnostic technique, they showed that a stiff main chain produced a disordered structure in the side chains and that a flexible main chain leads to a high degree of order in the side chains. This result indicates that the side chains are separated by such a distance that they are out of contact unless the main chain is badly distorted. Nerger *et al.* [248] studied polyesters and polyurethanes bearing long side chains. They claim a degree of order for one of their polyurethanes which leads to the resolution of three Bragg peaks, which is surprising as one would have thought that the way the side chains are spaced out along the main chain would lead to disorder in a manner similar to that of the polyamides referred to above. He *et al.* [249] studied poly[2,4-hexadiyne-1,6,diol *bis*(*p*-toluenesulphonate)] and succeeded in forming well ordered multilayers which lead to the observation of three Bragg peaks.

Polypyrroles have been extensively studied recently because their conjugated structure leads to electronic conductivity. It is thus not surprising

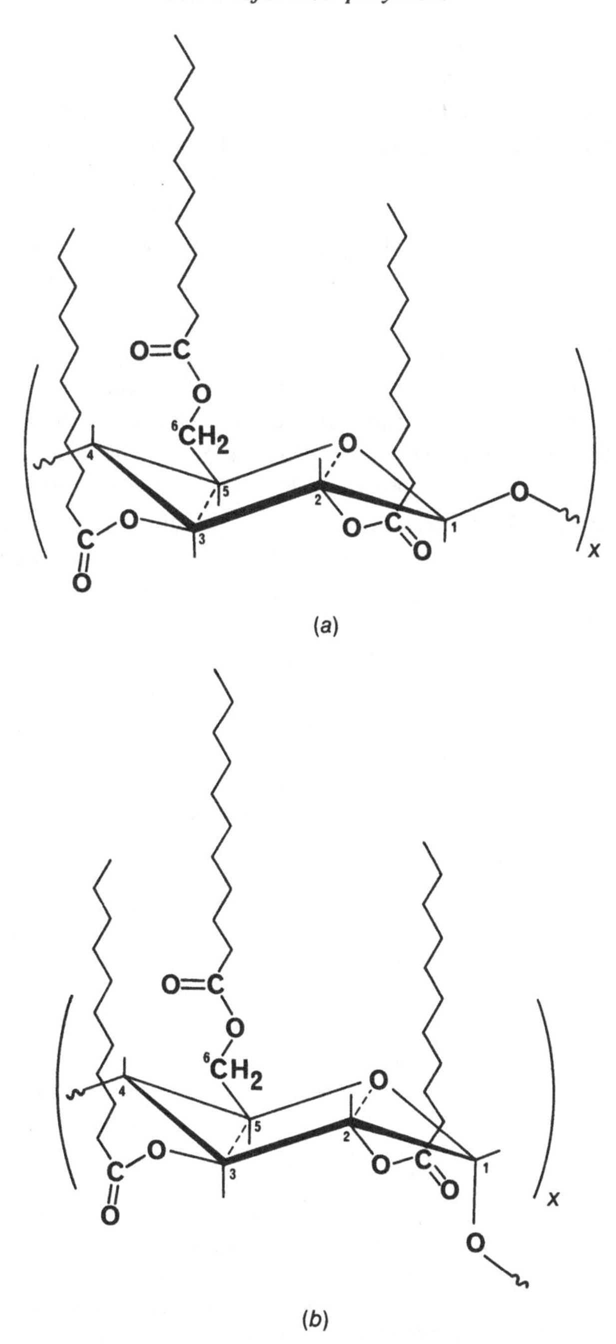

Figure 5.9. (*a*) A cellulose ester. The scale of the side chains is reduced for convenience. (*b*) An amylose ester. The vertical scale is again reduced.

that LB films of these materials have also been studied. The basic pyrrole structure is shown in Figure 5.10(*a*) and a possible conformation of a copolymer of pyrrole and a pyrrole derivative is shown in figure 5.10(*b*). The polymerisation process can be brought about either by the electrolysis of an existing multilayer of pyrrole derivative monomers deposited on a conductive substrate as was shown by Iyoda *et al.* [250] or at the air/water interface, the process being brought about by the presence of $FeCl_3$ dissolved in the subphase, as was shown by Yang *et al.* [251]. This latter procedure is only successful if a large excess of pyrrole, with respect to the pyrrole derivative employed, is used. It is not yet clear why this should be so, as the final product is formed from equimolecular quantities of the starting materials. Rotation is possible about the bonds joining the pyrrole residues but, if the chains are to be straight, the conformation shown in Figure 5.10(*b*) must exist. Such a conformation can be encouraged by using a combination of monomers such as is shown, for example, in this figure and which was employed by Yang *et al.* [251]. A number of other papers have appeared dealing

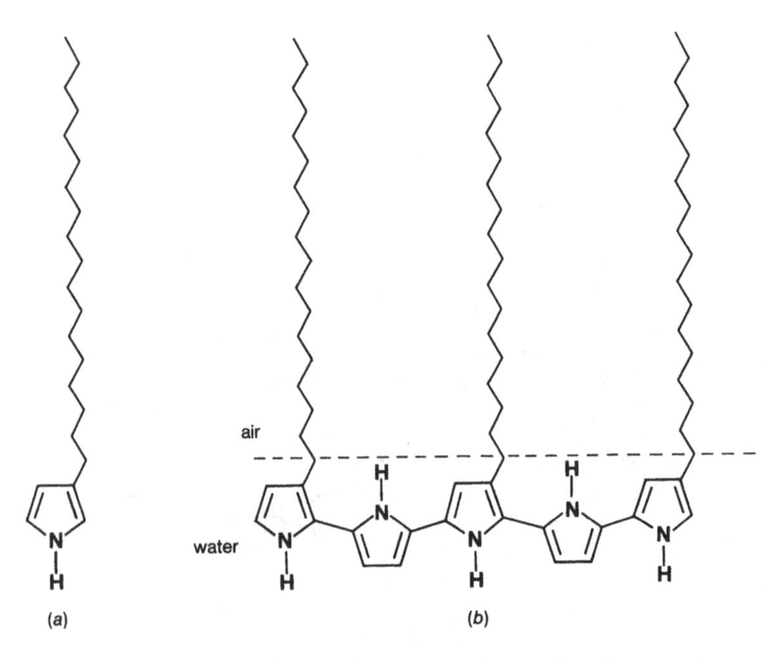

Figure 5.10. (*a*) A long chain pyrrole. (*b*) A polypyrrole. The pyrrole groups have a dipole moment and are hence hydrophilic. Alternation of the direction in which the pyrrole and long chain pyrrole groups are oriented allows the polymer chain to reside at the air/water interface without straining the structure.

with polypyrrole LB films but very little progress has been made in the characterisation of their structure.

Carr *et al.* [252] studied LB films made from preformed polymers based on the polysiloxane structure. Such materials have attracted attention as liquid crystal side chain polymers as they remain in a mesophase at low temperatures but, just for this reason, they seem unlikely materials to form stable LB films.

Tamura and his collaborators [253, 254] have investigated the effect of replacing hydrocarbon side chains by perfluorinated side chains in a

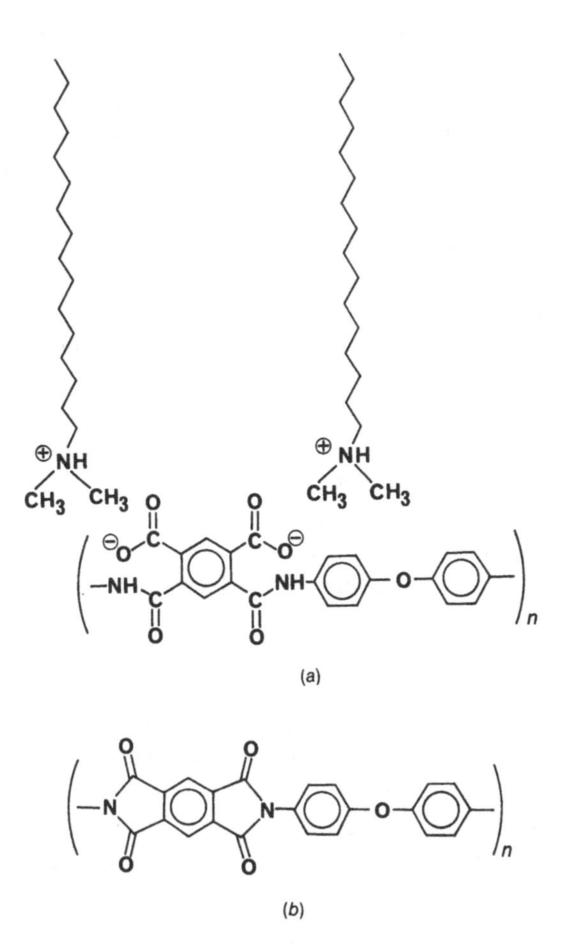

Figure 5.11. (*a*) A polyamic salt which will spread at the air/water interface and form LB films. (*b*) A polyimide which can be produced as a multilayer by heating a multilayer of the polyamic salt and driving off the long chains associated with the salt.

number of polymers used as LB materials. There seem to be various technological advantages in this procedure but very little has been published about the structure of the multilayers so formed.

In order to form a monolayer of a polymer at the air/water interface and to assemble multilayers of this material, one needs to employ a material which bears both hydrophilic groups and hydrophobic side chains, as we have already seen. This fact means that a considerable part of the volume of the multilayer must consist of side chains which have no other purpose than to facilitate the assembly of the film. Kakimoto *et al.* [255] devised an ingenious way of circumventing this difficulty. They synthesised a polyamic salt shown in Figure 5.11(*a*) which could be spread at the air/water interface. LB multilayers of this material were formed on a quartz plate, deposition being in the Z mode. These films were then immersed in a mixture of acetic anhydride, pyridine and benzene (1: 1: 3) for about 12 h, which removed the long side chains and produced the polyimide shown in Figure 5.11(*b*). Low angle X-ray diffraction patterns were obtained from the resultant multilayers which suggest a monolayer thickness of about 0.4 nm. This thickness multiplied by the number af layers deposited agrees well with the film thickness measured by a stylus device. A very similar technique has also been described by Uekita *et al.* [256] but, in this case, the removal of the side groups was brought about by heating the multilayer to about 400°C. An analogous procedure for the formation of poly(*p*-phenylene vinylene), illustrated in Figure 5.12, has been described by Nishikata *et al.* [257]. Here the initial polymer was a poly(sulphonium) salt which deposited in the Z mode and the side groups were driven off by heating to 200°C. Era *et al.* [258] carried out a similar procedure and published at about the same time.

Much subsequent work has been carried out on the formation and characterisation of polyimide multilayers ([259–63] and other papers of less relevance to the theme of this book). It has been established that there is substantial orientation of the polymer axes in the direction of dipping, an effect which increased with the length of the polymer chains. Thin films consisting of ten monolayers were far more defect-free than fatty acid films of comparable thickness. Progress has also been made

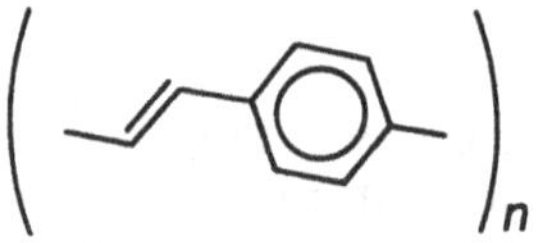

Figure 5.12. Poly (*p*-phenylene-vinylene).

in the preparation of highly anisotropic films of poly(*p*-phenylene vinylene) [264–6].

There remains one very curious feature of the polyimide films discussed above. Some of the side chains removed contain in excess of 16 carbon atoms and are therefore relatively massive. Neverthless, they appear to diffuse through a tightly packed polymer multilayer consisting of up to 200 monolayers. Undoubtedly this process takes place, but it is hard to understand the mechanism.

Other polymer films formed from alternate layers of two distinct materials will be discussed in Section 5.6

5.5. Rigid-rod polymers

The polymers used to form LB films which have been discussed so far have relatively flexible main chains. Thus, even though they are confined to the plane of the air/water interface, they can take on complex conformations and, in some cases, double back and cross over themselves. There is thus an attraction in the possibility of forming polymers which are constrained to remain in a conformation corresponding, at least approximately, to a straight line, but which have amphiphilic properties which ensure that this line is parallel to the water surface. Materials whose monomers are chiral, and also those which are isotactic, have a natural tendency to take on a helical structure. The amylose esters discussed above [243] probably behave in this manner. Many polypeptides are also examples of this type of polymer. These materials can form helices stabilised by hydrogen bonds as illustrated in Figure 5.13. Such conformations are examples of the famous α-helix first predicted by Pauling *et al.* [267]. Synthetic polyglutamates have been available commercially for some years now. Malcolm, in a long series of studies, has explored the behaviour of these materials at the water surface [268–72]. It appears that the ester bonds which attach the side groups have sufficient polarity to be hydrophilic whereas the main chain tends to be hydrophobic and the helix is thus stable at the air/water interface. Prior to this work it was believed that such polymers would unwind at the water surface. The polyglutamates most studied in this connection are poly(γ-benzyl-L-glutamate) and poly(γ-methyl-L-glutamate). Both exhibit isotherms of the type illustrated in Figure 5.14. The sharp rise in surface pressure and associated discontinuity correspond to collapse of the monolayer and formation of a bilayer, as was first suggested by Malcolm. Takenaka *et al.* [273] formed LB films of poly(γ-benzyl-L-

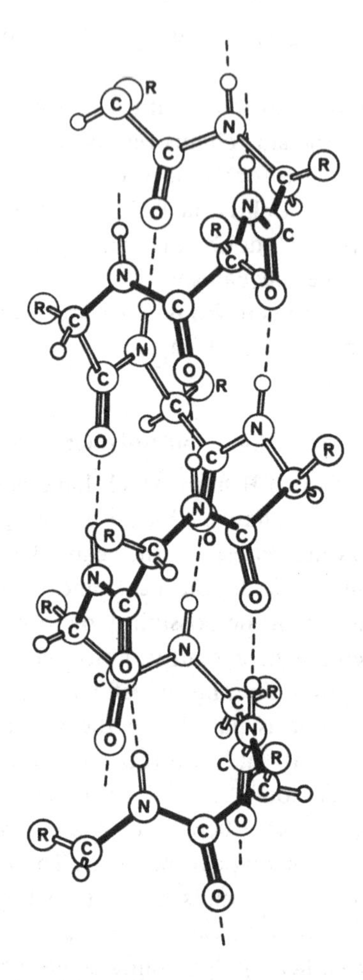

Figure 5.13. The α-helix. The side groups which characterise the particular amino acid residues of which the helix is formed are denoted by ®. The dotted lines indicate hydrogen bonds.

glutamate) at surface pressures of 5 mN m^{-1} and at 10 mN m^{-1} corresponding to the regions below and above the discontinuity and obtained deposition in the Z mode. Studies made by ATR and by transmission infrared spectroscopy show that the polymer axes reside at about 35° to the dipping direction. Dipping above 10 mN m^{-1} leads to a higher optical absorption for a given number of dipping cycles than does dipping at 5 mN m^{-1}. This result is consistent with the fact that two layers per cycle are deposited at the higher pressure but only one at the lower

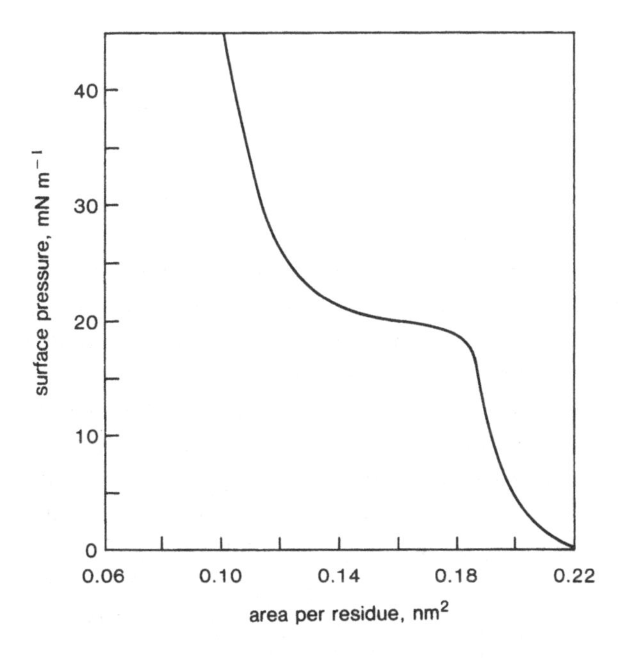

Figure 5.14. Isotherm obtained from poly(methyl-L-glutamate). The inflection is at the point at which a monolayer starts to collapse into a bilayer.

pressure. Similar results have been obtained by Takeda *et al.* [274] for poly(γ-methyl-L-glutamate). In addition Takeda *et al.* [275] studied the related material poly(ε-benzyloxycarbonyl-L-glutamate) and found that a series of discontinuities in the isotherm corresponded to the formation of a series of ever thicker multilayers at the water surface, which could be picked up by the LB technique and identified. The results concerning deposition of bilayers reported above for the two polyglutamates were confirmed by Winter and Tredgold [276], who formed multilayers above and below the transition points and measured the electrical capacity of such layers as a function of the number of dipping cycles. Jones and Tredgold [277] formed multilayers of poly(γ-benzyl-L-glutamate) using the horizontal touching method which does not influence the orientation of the polymer chains. They showed, using polarised infrared spectroscopy, that, in addition to the dipping process, the compression process influences the orientation of the polymer chains at the air/water interface. In the monolayer region, the chains are oriented perpendicular to the compression barrier, but they appear to be predominantly parallel to the barrier in the bilayer regime. Hickel *et al.* [278] have studied long alkyl chain substituted polyglutamates. They have succeeded in forming thick multilayers which can be used as optical waveguides and have obtained attenuations of between 2.5 and 5.5 dB cm^{-1}, which indicates a relatively high quality film.

Helical polymers, formed as they usually are from chiral monomers, are themselves chiral. If an attempt is made to stack them in a regular parallel array, there will be a tendency for successive layers to be gradually twisted round with respect to the original layer. It is possible that the chiral effect is responsible for the fact that the polyglutamates tend to deposit at an angle of 35° to the dipping direction. A study of a racemic mixture of one of these materials would clarify this point.

Another family of materials which exist as rod-like polymers which can be deposited as LB films are the cofacial phthalocyanine polymers more pedantically known as phthalocyaninato-polysiloxanes. These materials were first described by Joyner and Kenney [279] and a schematic diagram of this structure is given in Figure 5.15. The material which resides at the centre of the phthalocyanine ring is usually either silicon or germanium and the bridging atom is oxygen. It is possible to produce an analogous structure in which the centre atom is aluminium and the bridging atom is fluorine. These materials are usually formed at temperatures of the order 400°C. They can be formed by evaporation *in vacuo* and their structure has recently been studied by Fryer and Kenney [280]. They succeeded in forming small single crystals of these materials by these methods but regular ordered films of them have not been obtained this way. Wegner and his collaborators [281–3] have used

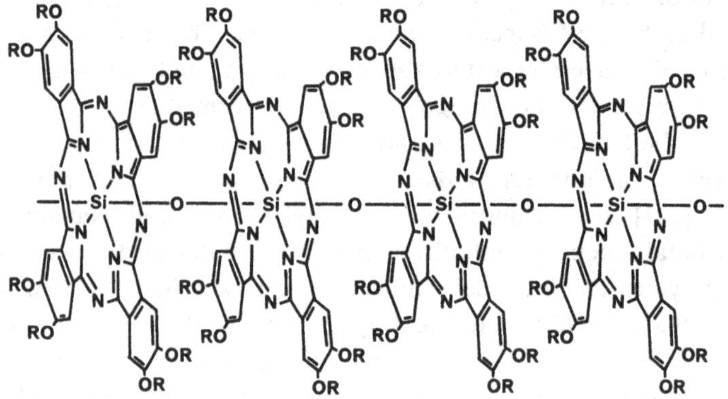

Figure 5.15. A schematic view of part of a bridge-stacked polyphthalocyanine.

an entirely different approach to synthesise analogues of these materials in which some of the peripheral hydrogens on the phthalocyanine molecules have been replaced by alkyl groups and the resulting polymers have been rendered soluble in ordinary organic solvents. Several of the materials so formed can be spread at the water surface and deposited as LB films, as was first shown by Orthmann and Wegner [282]. The structure of films of such materials has been explored by Rabe *et al.* [284] using the STM technique. Sauer *et al.* [285] have deposited good multilayers of these materials and have shown that the polymer chains tend to deposit in the dipping direction. They used polarised infrared spectroscopy to characterise their materials and obtained a dichroic ratio of 2.3. The influence of subphase conditions on the deposition of these materials has been studied by Kalachev *et al.* [286] and layer structure has been studied by Crockett *et al.* [287].

The families of materials discussed in this section have also been deposited from solution to form films having a thickness of the order 1 μm, orientational order being introduced by the application of a strong magnetic field. The diamagnetic nature of these materials is made use of, its effect being very much magnified by their liquid crystalline nature. This topic will be returned to in Chapter 7.

5.6 Alternating layers

The study of alternating or ABAB . . . film structures has attracted wide attention, as it has been seen as a road to the formation of practical devices which might compete favourably with similar devices formed in other ways. The two applications in question are the generation of second harmonic radiation (SHG) from a fundamental in the near infrared region and the detection of weak infrared signals using the pyroelectric effect. At the time of writing it has not proved possible by the LB technique to produce devices of either kind which show a clear superiority over devices formed by other less complex techniques. However, before turning to the problems of structure and order connected with ABAB . . . layers it is appropriate to give a brief discussion of the two proposed applications mentioned above.

In order to generate the second harmonic of an electromagnetic wave, one needs to make use of some device which has a non-linear property. In the case we are considering, the non-linear relationship made use of is that between applied electric field and electric polarisation. One can write

$$P=\epsilon_0(\chi_1 E+\chi_2 E^2+\chi_3 E^3+\ldots)$$

where higher orders in E make only trivial contributions to P. If we put $E=E_0\cos(\omega t)$ then the term in $\cos(2\omega t)$ in P is

$$\frac{1}{2}\epsilon_0\chi_2 E_0^2$$

Thus we seek a material having a large χ_2. Given that the term in χ_1 is much larger than the term in χ_3, a little reflection will show that non-zero χ_2 implies a material in which the increase of polarisation for a given value of positive E has a different magnitude from the increase of polarisation for an equivalent value of negative E. In other words the material must be non-centrosymmetric. This argument has been given in one dimension but it is trivial to extend it to three dimensions. The first requirement is thus molecules which have a non-centrosymmetric electric polarisability along the direction in which the electric field is to be applied. Suppose one is considering an LB film in which this direction is the normal to the plane of the film and that it has been deposited in Y mode. Then the two components of each bilayer will cancel out one another's contribution to P. Thus to obtain a multilayer which will lead to the generation of a second harmonic, one must either deposit alternate layers of materials having different electrical properties or one must deposit in the Z mode. Most attempts to produce second harmonic generators have used the first of these possibilities, but one highly successful device has been produced using the second possibility. It is also possible to apply the electric field in a direction tangential to the plane of the film and we will return to this concept below.

Attempts to produce pyroelectric devices have been less numerous than those aimed at producing second harmonic generators. Clearly a pyroelectric device must be non-centrosymmetric in character and must thus be formed either from an alternate layer structure or from a Z layer structure.

Many of the papers devoted to the two topics discussed above are essentially technological in character and do not attempt to characterise structure and order and thus are not really relevant to the subject matter of this book. However, a number of such papers have made important contributions to the field of non-linear optics and infrared detection. Thus the remainder of this action will consist of a discussion of those papers whose content is directly relevant to questions of order and structure followed by a catalogue of some of the other more important papers in this general field.

Returning to the topic of second harmonic generation, it should be borne in mind that the amplitude of the second harmonic is usually very much less than that of the fundamental. Thus, if one considers a fundamental ray propagating from left to right, the fundamental radiation reaching material in the right hand portion of the specimen will have a very much larger amplitude than the second harmonic radiation generated in the left hand part of the specimen and propagating through the right hand part. It is thus valid to neglect the effect of the second harmonic on the material in the right hand part so far as the generation of *its* second harmonic (which would be the fourth harmonic of the fundamental) is concerned. However, the second harmonic generated in the left hand region will propagate to the right and add to the second harmonic generated in the right hand region. If the phase velocity of the fundamental and the second harmonic are the same, there will thus be a simple additive effect, but if they are not there will be a gradual change in relative phase as between the locally generated second harmonic and the second harmonic propagated from further to the left. For a sufficiently thick specimen, these two components will eventually get out of phase and interfere in a destructive manner. To avoid this destructive interference, it is necessary to introduce some additional effect so that the phase velocity of the fundamental and that of the second harmonic are equal. For example the influence of wave-guiding on the velocities of the two waves can be made use of. Such procedures are known as phase matching. They do not concern us further here as we will only consider materials having a thickness up to a few wavelengths at most.

With the limitations discussed in the last paragraph, one can assume that the amplitude of the second harmonic will increase in a manner linearly proportional to the distance through which the fundamental wave has travelled. It should be noted that experiments are usually carried out with the direction of propagation at 45° to the plane of the film so that there is an appreciable component of the electric field normal to the plane and thus in a direction suitable for generation of a second harmonic wave in the manner so far envisaged. The finer points of the optics and geometry of this process are not germane to the theme of this book and will not be entered into here. However, it should be noted that the energy flux associated with the second harmonic is proportional to the square of the electric field and thus, if the other parameters are held constant, the second harmonic energy flux will be proportional to the square of the number of bilayers in the material. If a regular ABAB . . . structure has not been produced, this will not be the case. Thus a plot of the square root of the energy flux versus number of layers deposited is a sensitive indicator of the degree of perfection of the alternating layer structure.

We can now turn to an account of published work dealing with experiments in which attempts have been made to apply the concepts outlined above. The large number of papers dealing with SHG from *monolayers* of organic molecules deposited on solid substrates will not be discussed as, though many of them are important considered as optical studies, they have less relevance here. Girling *et al.* [288] were the first workers to produce an alternating layer structure containing a second harmonic generating molecule using the LB technique. They employed alternate layers of ω-tricosenoic acid and a merocyanine dye (see Figure 5.16(*a*)) and obtained the quadratic relationship between number of layers and second harmonic energy flux discussed above. However, they were only able to obtain this behaviour for the first three merocyanine layers, indicating that this particular alternating structure is not particularly stable. Furthermore, the dye itself is not very stable and needs to be exposed to an atmosphere containing ammonia. Thus these workers turned to the more stable hemicyanine (Figure. 5.16(*b*)) based materials

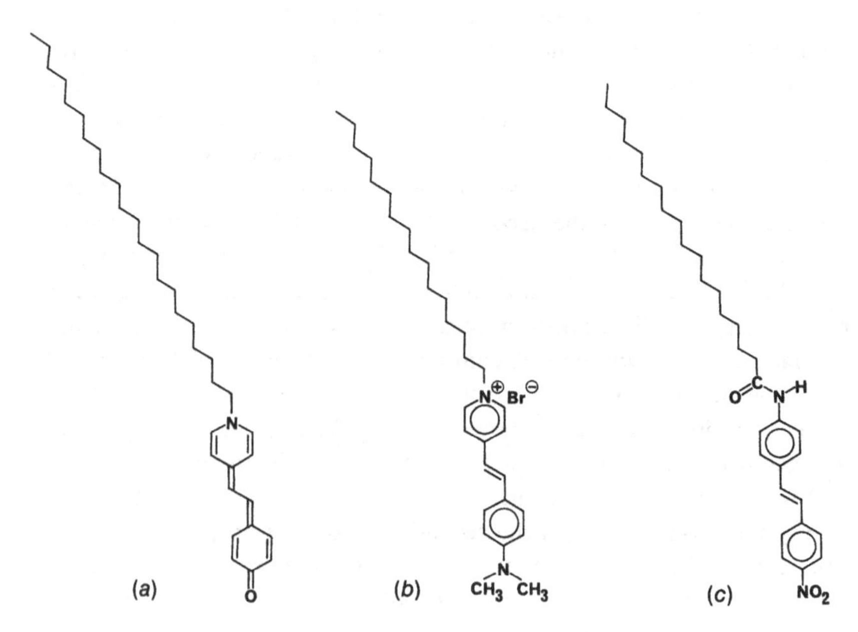

Figure 5.16. (*a*) A long chain merocyanine. (*b*) A long chain hemicyanine. (*c*) A long chain nitrostilbene.

in their subsequent work (Girling *et al.* [289]). Their initial study of this material showed little improvement on the results reported for merocyanine and, indeed, there is less tendency to follow a quadratic relationship between second harmonic energy flux and number of layers as compared with the merocyanine system. However, in their subsequent collaboration with the Durham group [290, 291] they were able to deposit alternate layers of their hemicyanine material and a nitrostilbene compound (Figure 5.16(*c*)) and were able to obtain the desired quadratic behaviour for up to ten bilayers of these two materials. These two materials deposited in this way are able to reinforce their individual optical effects. In a further study [292] they found that *dilution* of the hemicyanine layer by an optically inactive fatty acid actually increased the intensity of the second harmonic generated. The reason for this phenomenon is not entirely clear. Molecules having a hyperpolarisable character (that is to say, having an ability to generate a second harmonic component) also have a fixed dipole moment. If two such molecules are placed side by side, there will be a tendency for the electric field from each to depolarise the other and thus, possibly, reduce one another's hyperpolarisability. It is, however, surprising that this effect should outweigh the effect of simply reducing the quantity of hemicyanine present. Tredgold *et al.* [293] deposited alternate layers of the merocyanine dye discussed above (Figure 5.16(*a*)) and a preformed polymer and were able to obtain an increase of hyperpolarisability for up to 70 bilayers of these materials. However, even in this case, a true quadratic behaviour was not obtained, showing that a perfectly ordered alternating layer structure had not been obtained. Tredgold *et al.* [229] examined the effect of alternating an inactive *monomer* with a polymer bearing hemicyanine side chains. X-ray diffraction studies showed that a superlattice structure was obtained for this system. However, for a pair of polymers, one of which carried the hemicyanine group, it was impossible to demonstrate the existence of a superlattice.

Anderson *et al.* [294] introduced an interesting new concept. They used a repeat unit which can be represented as ABCC. A and B are polymers bearing hemicyanine side groups so arranged as to reinforce their optical effects when the polymers are alternated. C is a fatty acid and provides a skeletal structure which prevents the polymer multilayers becoming disordered. In their turn the polymer layers prevent the fatty acids growing in an epitaxial manner and thus forming large crystallites. They obtained good quadratic behaviour for up to ten polymer bilayers. Stroeve *et al.* [295] used this concept to stabilise Z layers of polymers bearing hemicyanine side chains and thus produced a non-centrosymmetric structure. Young *et al.* [296] attempted to extend this technique by introducing a cadmium arachidate bilayer after each ten polymer bilayers but obtained only a partial improvement in order as characterised by second harmonic studies.

Recently Cresswell *et al.* [297] have produced alternating layer structures which they have characterised by SHG and X-ray diffraction and have obtained a partial degree of order. Era *et al.* [298] have produced thick alternating films which exhibit quadratic dependence of second harmonic intensity on film thickness and which show a reasonable degree of order when studied by X-ray diffraction.

Ashwell *et al.* [299] have succeeded in forming good alternating layer structures of the materials shown in Figure 5.17 and have observed

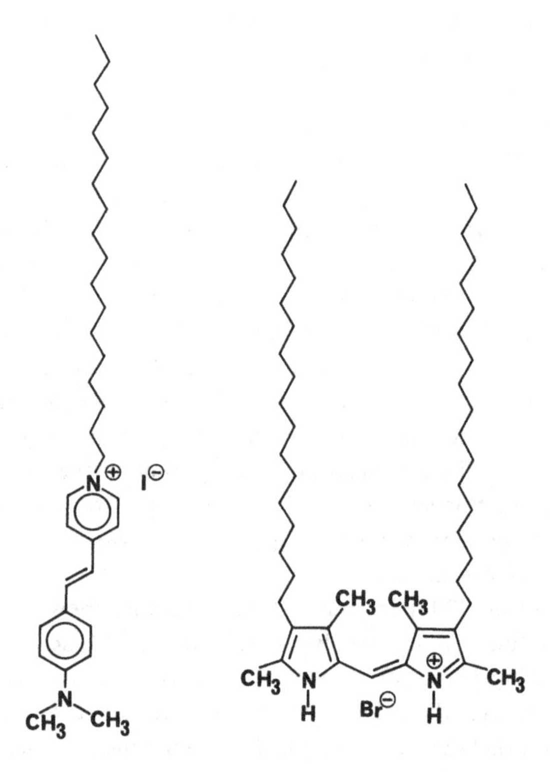

Figure 5.17. The two molecules discussed in [299]. These molecules can be deposited by the LB technique in alternate layers and form a non-centrosymmetric structure which can be used to generate a second harmonic in the visible spectrum. It is believed that the single hydrocarbon chain of one material interdigitates with the double chain of the other material, thus forming a stable structure.

quadratic behaviour for films having up to 200 layers. They make the very plausible suggestion that the long chain on 1 interdigitates with the two long chains on 2 and it is this fact that leads to the conservation of order over these long distances. At the time of writing, this supposition still has to be confirmed.

Other interesting papers in this general field are listed at the end of this section.

If thin LB films having substantial pyroelectric properties could be formed, they would have an important application as infrared image detectors. The essential requirement for pyroelectricity is a non-centrosymmetric structure. The first realisation of this idea was by Blinov *et al.* [300], who used a Z-type structure. In view of the instability of most Z-type structures, there would be obvious advantages in using instead an alternating structure, and the first attempt to do this was made by Smith *et al.* [301]. They used alternating layers of the merocyanine dye shown in Figure 5.16(*a*) and ω-tricosenoic acid and various other pairs of materials including long chain amines, and obtained a small but significant pyroelectric behaviour. Considerable effort has been devoted to finding materials which would lead to technically useful pyroelectric behaviour, but so far without success. A selective list of papers dealing with this topic is given at the end of this section.

In addition to those efforts to harness the concept of alternating LB films to technical applications several papers have been published dealing purely with the scientific aspects of this topic. Tredgold *et al.* [302] studied alternating layers of arachidic acid and two different amphiphilic porphyrins, mesoporphyrin IX dimethylester and mesoporphyrin IX diol copper complexes, the structures of which are shown in Figures 5.18(*a*) and (*b*) respectively. In the case of the diester, it was impossible to obtain a stable superlattice and low angle X-ray diffraction produced Bragg peaks which corresponded to Y layers of the fatty acid alone. On the other hand, the diol was capable of producing superlattices with both arachidic and stearic acids which gave good clear Bragg peaks corresponding to the expected superlattice spacing. Electron microscopy confirmed that the diester and arachidic acid produced an image of a multidomain structure, whereas the diol and arachidic acid produced a uniform and structureless image, as would be expected from the X-ray results. The diol is more hydrophilic than the diester and this must be the factor which causes the difference in behaviour between the two materials. As all these materials were deposited from good stable films at the air/water interface, it is evident that the diester molecules must be capable of diffusing *after* deposition for distances of up to 1 μm.

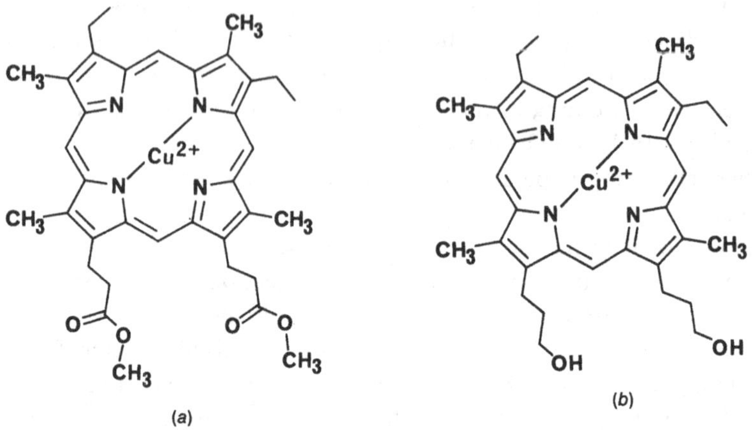

(a)

(b)

Figure 5.18. (*a*) Mesoporphyrin IX dimethyl ester. (*b*) Mesoporphyrin IX diol.

Okada *et al.* [303] deposited alternate layers of an amphiphilic diacetylene compound or arachidic acid and poly(isobutyl methacrylate) polymerised the former material after deposition by irradiation with ultraviolet light. They also studied sandwich structures in which alternate bilayers of the two materials were deposited and in which three layers of the polymer and two layers of the diacetylene material were deposited. X-ray diffraction and optical spectroscopy were used to characterise the resultant films. The simple alternate layers did not produce a superlattice peak, though superlattice peaks appeared at the expected places for the other stuctures studied. The implications of these results are not yet clear. Lvov *et al.* [304] studied structures which consisted of alternate *bilayers* of 4-n-octadecylphenol and barium behenate. They characterised these materials using low angle X-ray diffraction and by electron diffraction. A full structural analysis was undertaken and an electron density profile obtained which agreed well with the profile which one would expect from the structure and postulated packing of the molecules. Strong electron bombardment caused cross linking of the octadecylphenol molecules and reduced the amount of order in the films.

Additional papers dealing generally with alternate layer structures are referred to at the end of this section.

The obvious ways of producing non-centrosymmetic films are either to form Z layers or to form alternating layers of two distinct molecular species. In either case, the direction along which the phenomenon of interest will appear lies in or near the normal to the plane. However, another possibility exists. Suppose a Y structure of non-centrosymmetric molecules is deposited having a herringbone-like structure, then the hyperpolarisable portions of the molecules will reinforce their effect along a line which lies in the plane of the film. As such a structure is most likely to be created by the effects of dipping, this line will lie in or near the dipping direction. This ingenious concept was first put into effect by Decher *et al.* [305] using the molecule shown in Figure 5.19. Their excellent quadratic results are shown in Figure 5.20. The idea was extended by this group in a number of further papers, of which the most important are probably by Bosshard *et al.* [306, 307] and Pasquier *et al.* [308]. Further papers on this topic are listed at the end of this section.

A number of attempts to generate second harmonic waves using Z deposition have been made with only modest success; some of the more interesting papers on this topic are listed at the end of this section. However, Ashwell *et al.* [309] succeeded with the material shown in Figure 5.21 in obtaining quadratic behaviour for up to at least 50 layers. The stability of this structure is no doubt due to the fact that these molecules are zwitterions. (That is to say that one end of the main body of the molecule is positively charged and the other end negatively charged.) The resultant electrostatic forces thus serve to stabilise the Z structure at least in the direction of the molecular axes. There remains the problem of why the forces between molecules lying side by side do not destabilise the structure. Perhaps the long hydrocarbon group R helps to prevent such destabilisation.

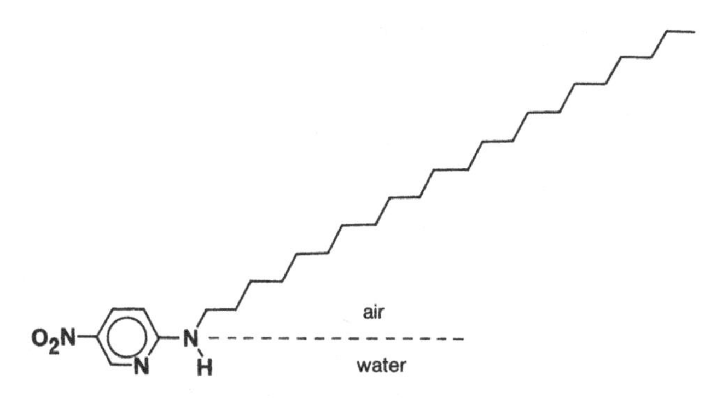

Figure 5.19. The structure of the material discussed in [305] (see also Figure 5.20). This material, when deposited by the LB technique, produces a non-centrosymmetric structure in which the dipole moment and hyperpolarisability are in a direction tangential to the plane of the multilayer.

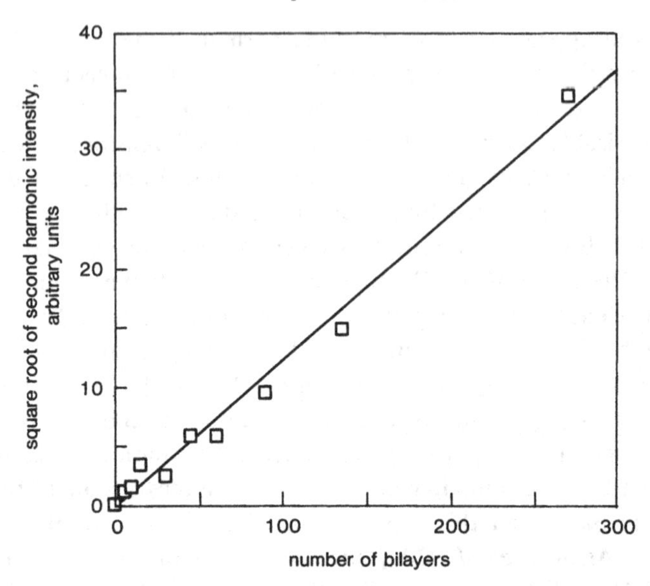

Figure 5.20. The square root of the second harmonic intensity versus number of layers obtained using the material shown in Figure 5.19. Reproduced from Decher, G., Tieke, B., Bosshard, C. and Guenter, P. 1988 *J. Chem. Soc: Chem. Commun.* 933–4. (Reproduced by kind permission of the authors and of the Royal Society of Chemistry.)

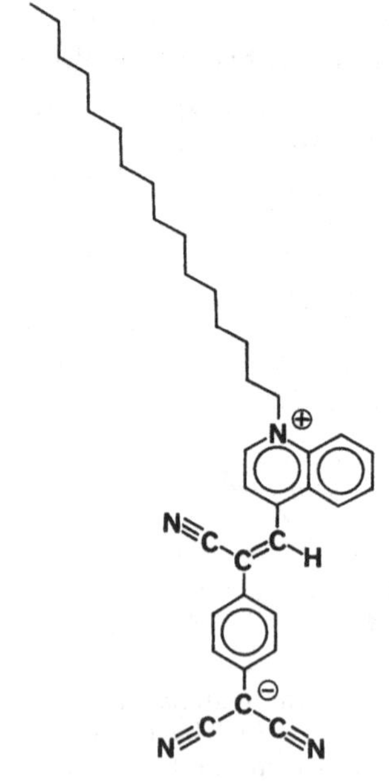

Figure 5.21. The material discussed in [309]. This substance deposits in the Z mode and produces layers which can generate a second harmonic at optical frequencies.

Further papers of interest dealing with:

SHG are [310-20]
pyroelectricity are [321-8]
structure are [259, 329-32]
non-centrosymmetric Y layers are [333-42]
Z type SHG are [343-50].

5.7 Conduction in the plane of the film

A substantial degree of effort has been devoted to attempting to produce conductive Langmuir–Blodget films, not without some success. Most of the systems employed have used TCNQ or variants of this compound together with some other material and have usually made use of iodine doping to obtain conductivity. In many cases some form of alternating structure has been employed and thus this topic might be treated in this chapter. However, the degree of order obtained is usually low and often ill-characterised. Thus, notwithstanding the interest of this field of study, it will not be discussed in this book.

Recently Japanese workers have succeeded in forming well ordered LB films from single compounds which show a high in-plane conductivity and whose layer structure has been characterised by low angle X-ray diffraction [351, 352].

6
Self-assembly

6.1 Introduction

The name self-assembly is an unfortunate one as it implies the achievement of something approaching the synthesis of artificial life. However, this term has now been generally accepted and so will be used here. It has two distinct but related meanings. The majority of papers bearing this phrase in their titles concern monolayers of organic molecules adsorbed on solid inorganic surfaces. These rather simple systems can be studied and characterised at a level of detail and rigour which it is difficult to achieve with the other systems discussed in this book. They can also be formed using very simple apparatus. For both these reasons they have made a strong appeal to surface chemists. There exists, however, a more limited group of papers in which treatment of the initial organic layer by a succession of reagents has made it possible to build up ordered multilayers. In principle, this latter technique should make it possible to form the analogues of Langmuir–Blodgett Z-type multilayers and thus use relatively simple chemical methods to construct non-centrosymmetric systems of use in technology, as discussed, for example, in Chapter 5. So far such applications of this technique have not proved practicable and the difficulties involved will be discussed later in this chapter.

6.2 Monolayers formed from carboxylic acids

In recent years study of the absorption of small molecules on well characterised single crystal surfaces has attracted many research workers. Here, however, we will only be concerned with relatively large molecules such as, for example, long chain fatty acids. In 1946 Bigelow *et al.* [16] showed that a carboxylic acid dissolved in a non-polar solvent will adsorb on to a hydrophilic surface immersed in this solvent and that, when this surface is withdrawn from the solvent, it will retain this adsorbed monolayer. The mechanism responsible for this behaviour is either the hydrophilic interaction discussed in Sections 1.3 and 1.4 of this book or a chemical interaction in which a proton is transferred from the acid to an oxygen atom on the substrate. The hydrophilic surfaces employed have included glass, mica or a metal bearing a thin oxide layer. A complete review of work in this field would be impossible to give here and would, in any case, be irrelevant to the problem of order in films. Reference will be made to a few particularly important papers. Early efforts to characterise such monolayers made use of reflection electron diffraction, which is a rather inappropriate technique to use in this context. In 1957 Chapman and Tabor [353] formed monolayers of long chain fatty acids on thin films of silver, copper, iron and cadmium deposited on glass microscope slides. The metal films were then floated off the slides together with the organic monolayers which had been deposited on them and were picked up on electron microscope grids. It was thus possible to make use of transmission electron diffraction to study the film structure. The molecular spacings were roughly what one would expect from a fairly closely packed array of nearly vertical hydrocarbon chains, though the rather precise values of spacing quoted by these authors seem, when viewed in retrospect, to overestimate the precision obtainable from this kind of experiment. All the metals used in this study which lead to successful diffraction experiments are now known to form uneven oxide when exposed to air and this fact must be taken into account when assessing these results. Nevertheless, these series of experiments were important in establishing that the structures of the absorbed monolayers were of the general kind proposed by Bigelow *et al.* [16].

In 1985 Allara and Nuzzo [354, 355] published the results of an extensive investigation in which adsorption took place on to an aluminium oxide layer formed on a film of aluminium deposited *in vacuo* on to a silicon wafer. Various carboxylic acids were dissolved in high purity hexadecane and allowed to adsorb from this solution on to the prepared aluminium oxide surface. The monolayers so formed were examined by ellipsometry and infrared spectroscopy. Contact angle measurements were made on the monolayer surfaces and radioactive labelled (tritiated) compounds were employed to study the interchange of adsorbed molecules with those in solution. Various other techniques of less immediate relevance to our present interests were also employed and reference to these two papers should be made for further particulars. Aluminium oxide is known to produce an exceptionally smooth and defect-free layer when a newly formed aluminium surface is oxidised by exposure to pure oxygen and thus the substrates on which the adsorbed layers formed were likely to be superior to those used in the previous study discussed. Adsorption took place either on a newly formed surface or on surfaces previously treated with dilute acetic acid.

It was found that, for carboxylic acids containing 12 or more carbon atoms, ellipsometry data indicated a film thickness which would be expected from nearly vertical orientation of the hydrocarbon chains and relatively tight packing. For shorter chain lengths it was not possible to form stable monolayers. It was shown that the kinetic processes involved in layer formation can take up to several days. Infrared studies lead to three important conclusions.

1. The carboxylic groups were bonded to the aluminium oxide by the transfer of a proton to the oxide substrate.
2. There was a randomness in the arrangement of these bonds which might indeed be expected from the amorphous nature of the substrate.
3. The hydrocarbon chains tilted at an angle of about 10° with respect to the normal to the substrate plane.

Chen and Frank [356] studied the adsorption of a series of fatty acids on to both aluminium oxide and clean glass. In particular they studied the variation of appropriate peaks in the infrared absorption spectra as a function of time in order to monitor the kinetics of adsorption. They also studied the variation of contact angle of both water and hexadecane with time allowed for adsorption.

Turning first to the infrared results, one assumes that the infrared absorbance is proportional to the coverage. These authors assumed that the absorption was described by the equation

$$\frac{\mathrm{d}\theta}{\mathrm{d}t}=\frac{k_a}{N_0}c(1-\theta)-\frac{k_d}{N_0}\theta$$

where

θ is the fractional coverage
k_a is the absorbtion rate constant
k_d is the desorption rate constant
t is the adsorption time
N_0 is the surface absorbate concentration at full coverage
c is the solution concentration.

Here the first term on the right hand side represents the rate of absorption and the second term represents the rate of desorption. This equation would describe the adsorption process very well if interactions between adsorbed molecules could be ignored. In fact, curves obtained by the integration of this equation can be fitted very well to the experimental results and can be used to make estimates of the energies of adsorption and desorption. For stearic acid (C_{18}) these are $12kT$ and $28kT$ on aluminium and $18kT$ and $31kT$ on glass respectively. The relatively large value of the energy of adsorption is attributed by these authors to the change in entropy involved, as the chains have to straighten out on adsorption. It is surprising that the rather simplistic theory used here works as well as it does.

The contact angle studies made by Chen and Frank [356] also deserve further study both in their own right and because contact angle measurements are routinely made by workers in this field. This topic is briefly discussed in Section 1.4. A drop of an ideal hydrophilic liquid (water) on an ideal hydrophilic surface should, and generally does, lead to a contact angle of zero. An ideal hydrophilic liquid on an ideal hydrophobic surface should lead to a contact angle of 180° but in fact a lesser angle is nearly always recorded. The converse situations are also true but an ideal hydrophobic surface is hard to define. Thus, for example, a surface consisting of closely packed methyl groups does not afford a good 'fit' to hexadecane which normally exhibits a contact angle of about 47° on such a surface. How far this behaviour can be explained in terms of internal energy and how far in terms of entropy is not clear. In view of these considerations, these authors have been wise to use contact angles 'as a qualitative indicator' only. In the opinion of the present author, other workers in this general field have relied too much on contact angle measurements and have attempted to squeeze too much information from them. Chen and Frank found that fatty acid monolayers were easily lifted off glass during contact angle measurements but not off aluminium oxide. This result is not in accord with the desorption energies quoted above and suggests that, perhaps, these energies are only roughly correct.

All the work discussed in this section involves rather ill-defined substrates and one is on firmer ground when discussing the adsorption of thiols on single crystal gold in Section 6.4.

6.3 The use of octadecyltrichlorosilane

In 1980 Sagiv [357] pioneered the use of octadecyltrichlorosilane, hereafter referred to as OTS, in the formation of self-assembled monolayers. Subsequently Netzer and Sagiv [17] extended the use of this type of material to the formation of multilayers and this work will be discussed in Section 6.5.

OTS was dissolved in a solvent consisting of 8% $CHCl_3$ 12% CCl_4 and 80% hexadecane. Clean glass slides were slowly dipped into freshly prepared OTS solution and the reaction shown schematically in Figure 6.1 took place. As soon as the hydrolysis takes place, polymerisation becomes possible and thus the initial solution has to be kept free of water. However, the water needed to bring about the first reaction shown in Figure 6.1 is assumed to be an adsorbed layer on the glass surface. Thus the first process to take place is physisorption followed by hydrolysis. The OH groups now attach to the Si atoms in the glass surface to which the water was previously bound. The polymerisation process finally takes place with elimination of water molecules. The precise nature of these reactions is somewhat speculative but the final result has been confirmed by various diagnostic techniques. Contact angle measurements using water and also hexadecane confirm that the final surface produced by this process consists of closely packed methyl groups. Subsequently Sagiv's group examined these monolayers using various other techniques [358, 359].

The treatment of X-ray diffraction given in Sections 2.3 and 2.4 assumes that a large number of layers of material exist. Most standard texts on X-ray diffraction extend diffraction theory to the case where only

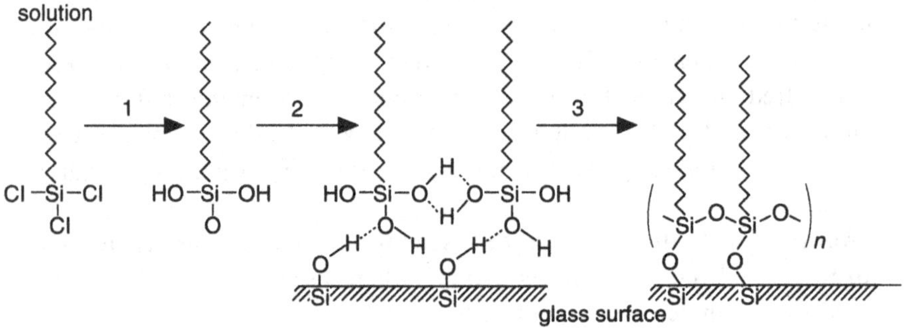

Figure 6.1. The use of OTS in the formation of self-assembled monolayers. The successive processes which take place are: (1) hydrolysis, (2) adsorption and (3) polymerisation with the elimination of water. Further details are given in the text.

a few layers are used and the diffraction phenomenon arising from the finite width of the multilayer is important. In such cases a number of subsidiary maxima occur and a full analysis makes it possible to deduce the structure of systems consisting of only a few monolayers. Pomerantz and Segmuller [38] applied this technique to relatively thin LB multilayers and subsequently collaborated with Sagiv's group [360] to apply it to films formed by self-assembly. They studied monolayers formed from OTS and concluded that the coverage obtained was better than 93% and could approach 100% in some cases. They also studied trilayers of this material formed by self-assembly and found much less satisfactory results in this case. This topic will be returned to in Section 6.5.

Cohen *et al.* [361] studied, amongst other things, the influence of heat on monolayers of OTS and of arachidic acid adsorbed on surfaces of aluminium oxide formed on aluminium surfaces. They used wetting studies and infrared spectroscopy and concluded that, whereas arachidic acid layers deteriorated in an irreversible manner at about 100°C, OTS layers survived intact to about 140°C. This result is attributed by these authors to polymerisation of the latter material but could also be due to a rather different bonding mechanism between the organic material and the aluminium oxide.

The difficulty in forming epitaxial layers of materials which deposit best on oxides is that there are very few oxides which are available in suitable single crystal form and which do not dissolve in the solvents employed. Mica seems a likely material and two attempts have been made to form OTS monolayers on this material [362, 363]. However, in neither of these studies was there clear evidence of epitaxy. Ogawa *et al.* [364] studied the self-assembly of nonadecenyltrichlorosilane on sputtered SnO_2 and found that, for films deposited at a temperature of 80°C there was a tilt angle of 32°. It is not clear from their paper whether these tilted regions were highly localised or whether they extended over macroscopic distances. Their principal diagnostic tool was ESCA, otherwise known as XPS. Here the surface is bombarded by monochromatic X-rays which excite electrons from inner shells of the constituent atoms of surface molecules whose energy levels are influenced by the valence state of the atoms. However, as the electrons so excited may have a mean free path of 4 or 5 nm, the results are not easy to interpret. The whole question of tilt will be returned to in Section 6.4.

The materials so far discussed in this section have all consisted of a headgroup and a long hydrocarbon chain. With a view to exploring the possibility of technical applications of self-assembly, Ulman and his

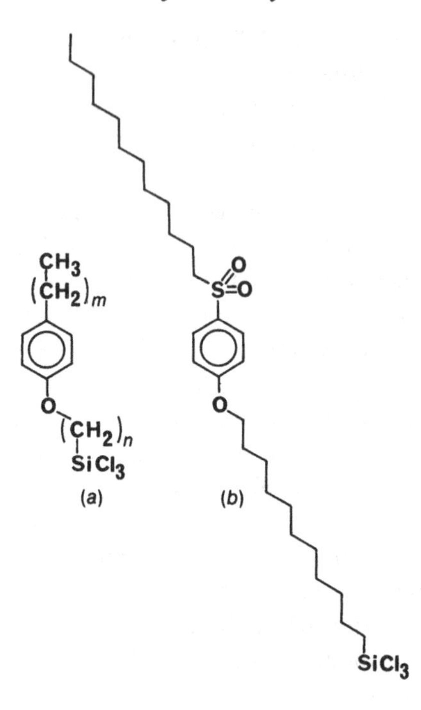

Figure 6.2. Trichlorosilanes containing an aromatic group which have been used for the formation of self-assembled monolayers.

collaborators [365, 366] have explored the self-assembly of trichlorosilane derivatives having aromatic functional groups at some point in the chain. Typical members of the two groups of materials studied are shown in Figure 6.2. Turning first to the material shown in Figure 6.2(*a*), it was found that, with $m=8$ and $n=11$, a monolayer was formed which was comparable in quality with layers obtained using OTS. However, for lesser values of m and n the layers formed, as judged by ellipsometry, infrared spectroscopy and wetting studies, were inferior in quality to OTS layers. Turning now to the material shown in Figure 6.2(*b*), infrared studies showed that self-assembled monolayers made from this material had a chain tilt of about 30° which is, according to these authors, explained by the bulky nature of the sulphone substituted aromatic group. However, as has been pointed out above, the factors influencing tilt are probably more complex than those arising from packing considerations.

6.4 Thiols on gold

In 1983 Nuzzo and Allara [367] showed that certain sulphur compounds had a strong affinity for gold and would bind strongly to a gold surface. The actual materials which they used have since been shown to be less interesting in this particular context than the related thiols, but this seminal paper has, nevertheless, led to a considerable interest and activity in the study of self-assembled monolayers of sulphur compounds formed on gold surfaces. Porter *et al.* [368] carried out a systematic study of n-alkyl thiols, $CH_3(CH_2)_nSH$, where $n=1$, 3, 5, 7, 9, 11, 15, 17 and 21 adsorbed on to gold from dilute solutions of these materials in hexadecane or ethanol. They characterised the resultant films by ellipsometry, infrared studies and electrochemical studies. For $n=11$ or greater, closely packed layers are obtained with a tilt angle of between 20° and 30°. For n less than 11, there is a gradual deterioration of film quality with decreasing values of n. Layers having large values of n were shown by electrochemical studies to be largely free of pin holes. It is generally believed that the adsorption process takes place by elimination of the terminal hydrogen atom and that the thiol is bound to the gold by a true valence bond. Furthermore, as is discussed below, it is possible to form good single crystal surfaces of gold by evaporation *in vacuo* on to other single crystals. Thus the way has been opened up to study the possibility of forming epitaxial layers of organic materials on an inorganic substrate.

Pashley [369, 370] showed that it is possible to evaporate gold *in vacuo* on to cleaved surfaces of alkali halides and hence form single crystal surfaces of gold. This method was adopted by Strong and Whitesides [371] who used cleaved surfaces of NaCl to produce 100 gold surfaces and BaF_2 to produce 111 surfaces. They adsorbed docosyl thiol, $CH_3(CH_2)_{21}SH$, and several other sulphur compounds on to these surfaces and studied them by electron diffraction. The gold films were about 80 nm thick and, when they had been formed, the alkali halide substrate was dissolved off in pure water. Various cleaning procedures were employed, for a description of which the reader is referred to the original paper, and the films were then immersed in a solution of the thiol in hexadecane. The films were then mounted on electron microscope grids and studied by transmission electron diffraction. The substrate and the thiol monolayers both produced diffraction patterns which could be simultaneously recorded. As is the case in the study of other thin organic films by electron diffraction, the film deteriorates rapidly on exposure to the electron beam and thus a focus was obtained and the beam was then quickly transferred to an adjacent region and the diffraction pattern recorded in a few seconds. It is not possible here to give a full account of this interesting and important paper and we will limit our discussion to the behaviour of docosyl thiol adsorbed on the 111 surface of gold. Such a surface has, of course, a hexagonal structure and the adsorbed thiol layer also has this symmetry which extends over distances comparable to the beam width (about 20 μm). However, there is no apparent translational epitaxy. Strong and Whitesides [371] suggest that there is a repeat unit corresponding to 7×7 atoms in the substrate where the atoms are counted along two principal lattice directions which are, of course, in this case, at 60° to one another. However Ulman [197] presents convincing arguments to support a different model. If one imagines a lattice formed from next nearest neighbours in the two-dimensional close packed hexagonal structure, one arrives at both the correct symmetry and about the correct lattice constant to account for the results obtained by Strong and Whitesides [371]. Ulman now supposes that this lattice is translated so that the thiol molecules are located at the centre of triangles formed by sets of three gold atoms. The postulated structure is shown in Figure 6.3. In support of this model he also quotes an as yet unpublished study made by H. Sellers, A. Ulman, Y. Shnidman and J.E. Eilers in which a refined Hartree–Fock calculation was made of the ground state of CH_3S species adsorbed on the 111

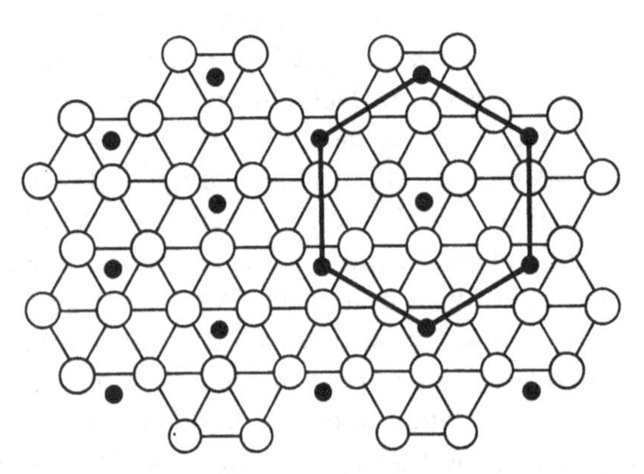

Figure 6.3. The postulated adsorption pattern of an alkanethiol on the 111 surface of gold. (From *Ultrathin Organic Films* by Ulman, A. 1991 (Academic Press, Boston). Reproduced by kind permission of the author and the publishers.) The open circles represent the sites of the gold atoms and the small solid circles the sites of the sulphur atoms.

surface of gold and the energy was minimised with respect to the lattice position of the CH_3S species. They found that the minimum energy corresponded to the configuration just described.

Strong and Whitesides [371] studied the variation of optical density of diffraction spots with respect to both radial and angular position and concluded that coherent domains of thiol were approximately 6 nm in diameter. They also used ellipsometry, and a relatively complex technique which depends on beam tilting to estimate the film thickness. Hence they deduce a tilt angle of about 30°. Chidsey *et al.* [372] used the diffraction of helium atoms originating at a temperature of 30 K to study a similar system and their results, though far less extensive than those of Strong and Whitesides [371], confirm their results. In this technique the collisions of the helium molecules passing through a narrow orifice have the effect of producing a very nearly monochromatic beam, the low energy of which does little to disturb the layer under study but which corresponds to a wavelength suitable for these diffraction experiments.

To recapitulate, the docosyl thiol monolayer adsorbed on the 111 surface of gold has a long range angular order but quite a short range radial order and a chain tilt of about 30°. Ulman *et al.* [373] put forward an interesting theoretical model based on chain to chain London forces and interchain hard core repulsions and conclude that the experimental results given above can be explained in terms of tightly packed tilted molecules. However, the reader will note that the monolayer in question has many of the properties of a hexatic structure and it seems to the present writer to be at least as likely that this system could be another example of the hexatic phase discussed in Sections 3.4 and 4.2. At the moment there is not sufficient evidence to decide the issue.

Both Langmuir–Blodgett and self-assembled monolayers have been studied by various workers by infrared techniques with a view to finding out how far the chain structure becomes disordered as the temperature is increased. In general, as indeed one might expect, an all-*trans* structure (one in which the chains are free of kinks) predominates in most such systems at room temperature but, as the melting point is approached, the number of *gauche* defects increases. Of particular interest in this context are the studies made by Nuzzo and his collaborators [374, 375], who examined the behaviour of the C_{16}, C_{17} and C_{22} thiols. They showed that such *gauche* defects as exist are concentrated near the free ends of the chains, a result which, though expected on common sense grounds, is interesting to have confirmed. These authors studied the materials in question at temperatures between 80 and 420 K, and found that, unlike bulk hydrocarbons, they do not exhibit any sharp phase transitions in this temperature range but rather a gradual change to a progressively more ordered state as the temperature is lowered. At the lowest temperatures employed there was evidence of spectroscopic band splitting, indicative of the formation of a phase having two molecules per unit cell.

Laibinis *et al.* [376] made an extensive study of n-alkanethiols deposited on gold, silver or copper. Owing to the reactivity of copper, it was only possible to obtain useful results with this metal if freshly formed metal surfaces were handled in an argon atmosphere and even then such results were not always reproducible. Their infrared results showed that the tilt angle of thiols on silver was about 12°, compared with about 30° for gold. A similar result was obtained by Walczak *et al.* [377]. Both groups of authors attribute these results to the different nature of the bonding of sulphur to silver as compared with gold and put forward qualitative but convincing arguments to support this point of view. Figure 6.4, which is copied from [376], illustrates the supposed bonding and also demonstrates why materials containing different numbers of methylene groups exhibit different surface and hence wetting properties. Laibinis *et al.* [376] carried out extensive modelling in an effort to explain their infrared results. They make a convincing case for the structure shown in Figure 6.5, in which there are two molecules per unit cell twisted about axes corresponding to the tilt direction but in opposite directions. Fenter *et al.* [378] studied the C_{20} thiol adsorbed on silver. They used helium atom diffraction and also the diffraction of X-rays impinging at an angle of 1.2° with respect to the tangent to the surface. They then studied the variation of density of diffracted radiation with respect to the *azimuthal* angle. The technique is equivalent to that

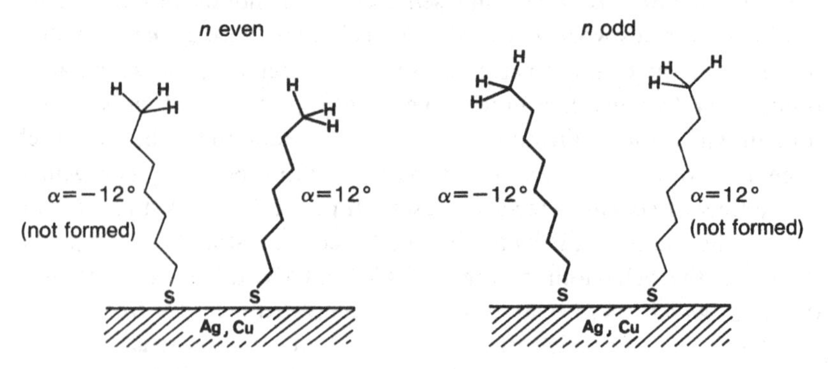

Figure 6.4. Illustration of the canted structure formed upon adsorption of the n-alkanethiols, $CH_3(CH_2)_nSH$, on copper and silver.

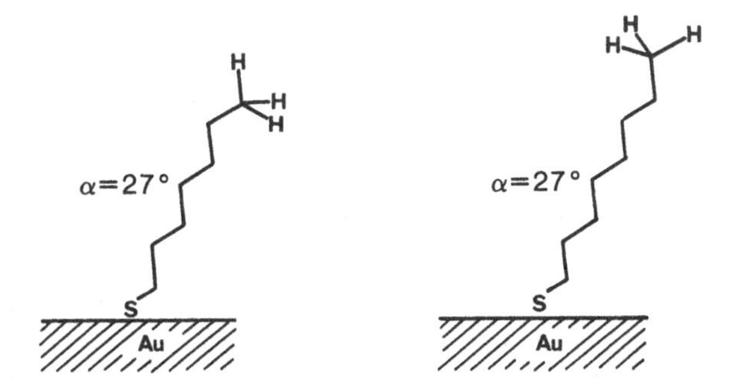

Figure 6.5. As for Figure 6.4 but for adsorption on gold. The FTIR spectra associated with the terminal methyl group and obtained with the electric vector normal to the surface alternate in amplitude as successive methylene groups are added to the chain. In the case of silver and copper this alternation is not observed. The authors explain this anomaly by supposing that the structures labelled 'not formed' in the diagrams do not occur. These results imply that the number of methylene groups do not influence the adsorption process in the case of gold but do so in the cases of silver and copper. Figures 6.4 and 6.5 are printed by kind permission of the authors and the American Chemical Society from Laibinis, P.E., Whitesides, G.M., Allara, D.L., Tao, Y.T., Parikh, A.N. and Nuzzo, R.G. 1991 *J. Am. Chem. Soc.* **113** 7152–676.

described in Section 3.3 and used there to study monolayers at the air/water interface. They obtained several interesting results.

(a) The nearest neighbour distance was between 0.467 and 0.477 nm as compared with about 0.497 nm for layers on gold.

(b) The X-ray diffraction peaks correspond to a number of domains rotated by plus or minus 18° with respect to the substrate lattice. They put forward a possible coverage arrangement which could account for such an angular displacement.

(c) Helium diffraction, which explores the outer ends of the molecules, leads to a coherence length of about 1 nm whereas X-ray diffraction, which explores the mean position of whole molecules, leads to a coherence length of about 12 nm. These authors claim that this discrepancy is accounted for by the fact that the outer ends of the molecules can move relatively freely as compared with the inner ends. It does, however, seem rather surprising that there should be a ratio of one order of magnitude between these two observed coherence lengths and part of this large ratio may arise from the nature of the two different experiments.

Bryant and Pemberton [379, 380] have employed surface Raman scattering to study thiol films on silver and gold and their results confirm the postulated difference in bonding.

The strong bond formed between thiols and gold and silver surfaces allows the possibility of forming molecules which have a wide variety of different functional groups at the opposite end to which the thiol group is attached, and thus of coating a noble metal surface with molecules bearing such functional groups. It is also possible to form surfaces bearing mixtures of functional groups. Such systems have provided useful tools for the use of surface chemists, and electrochemists but their study would lead to a sharp deviation from the theme of this book. It does, however, seem helpful to list a few such papers, particularly papers the contents of which bear, at least obliquely, on matters of order and structure. It should be understood, however, that many important and otherwise interesting papers concerning the adsorbtion of thiols on gold and silver have not been referred to. The author has found [381–91] of particular interest.

6.5 Self-assembly: multilayers

If the self-assembly technique could be applied successfully to the formation of ordered multilayers, it could replace the Langmuir–Blodgett method and avoid all the technical problems associated with troughs and deposition devices. We will see in this section that considerable progress has been made in this direction, but that most of the processes tried fail after the deposition of about ten layers, with one process having yielded about 25 layers on one occasion! The methods so far proposed all depend on successive immersion of the substrate in two or more reagents in a cyclical manner so that, after each complete process, one more layer is bound to the previous one. If we designate the proportion of the molecules in the first layer which have reacted to form a second layer when equilibrium is reached by α and the number of molecules in layer 1 by N_1, then the number in layer n will be given by

$$N_n = N_1 \alpha^{n-1}$$

Clearly if $\alpha = 1$ (or, in traditional chemical terminology, if the reaction has a 100% yield) the layers will build up in a satisfactory manner but, for all realistic lesser values of α, the numbers of molecules successfully deposited in a layer will decrease with n. For values of α very nearly equal to unity and fairly thin multilayers, reasonable success is

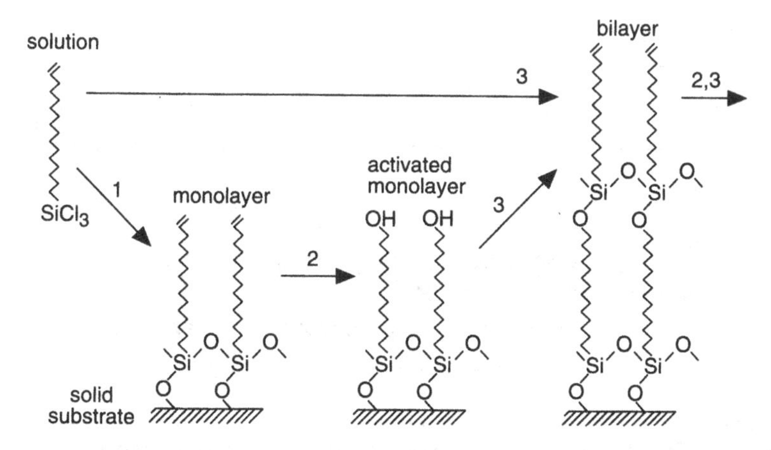

Figure 6.6. The use of OTS derivatives to form multilayers according to [17]. The various processes are represented by the following. (1) A trichlorosilane in a solvent consisting of 8% $CHCl_3$, 12% CCl_4 and 80% n-hexadecane from which the trichlorosilane chemisorbs. This reaction requires water, which is thought to come from an adsorbed layer on the substrate. (Various possible substrates are used including glass). (2) Chemical activation, which is brought about in two stages: (i) treatment for 2 min at 25°C by 1 M B_2H_6 in THF, and (ii) treatment by 0.1 M NaOH in 30% H_2O_2 for 1 min. (3) A further stage of chemisorption.

possible, but it is clear why multilayers corresponding to large values of *n* have yet to be produced.

The first successful attempt to form multilayers by self-assembly was made by Netzer and Sagiv [17], who formed a layer of OTS, modified so as to have a double bond at the extremity. They then used the reactions indicated in Figure 6.6 to produce an OH group at the terminal position. It was then possible to react a further layer of molecules bearing the trisilane group and thus, at least in principle, a multilayer could be formed. X-ray diffraction studies [360] showed, however, that, even for as little as three layers, the structure obtained was far from perfect.

Lee *et al.* [392, 393] devised a self-assembly process which makes use of 1,10-decane-*bis*-phosphonic acid and $ZrOCl_2$. After an initial step, shown in Figure 6.7, the surface is exposed alternately to aqueous solutions of these two reagents. The surfaces were washed between reactions. Figure 6.8 shows the film thickness as measured by ellipsometry as a function of the number of layers. It is evident that the process is successful, but it would not be suitable for the formation of non-centrosymmetric layers (but see [395] below). One obvious problem is

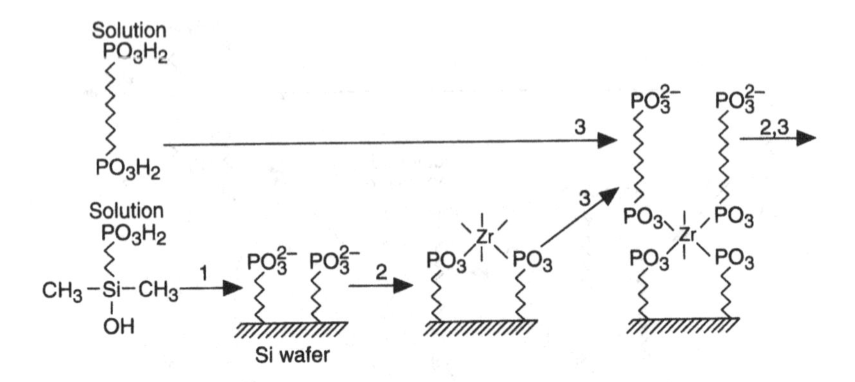

Figure 6.7. Multilayer formation by chemisorption using a zirconium salt according to [392]. The substrate is a silicon wafer. The successive processes are as follows. (1) Treatment of the substrate with a warm aqueous solution of a silanol. (2) After washing with water, the surface is exposed to a 5 mM aqueous solution of $ZrOCl_2$ and then washed. (3) Treatment by a 1.25 mM aqueous solution of 1,10-decane-*bis*-phosphonic acid.

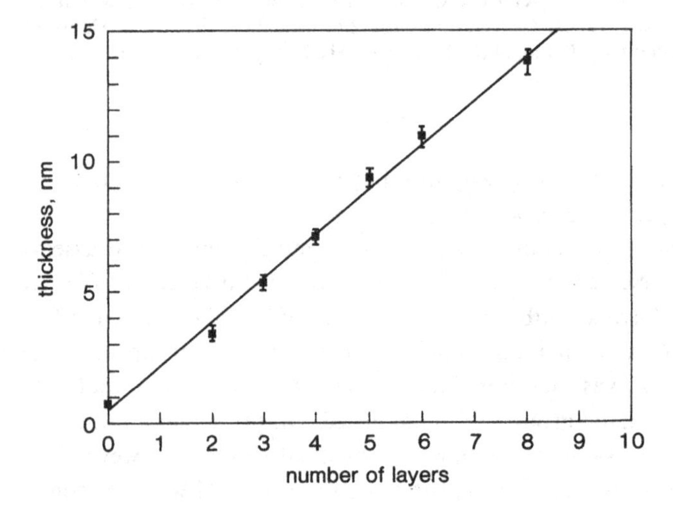

Figure 6.8. The thickness of films formed by the method shown in Figure 6.7 as measured by ellipsometry, plotted versus the number of layers. (Reproduced from Lee, H., Kepley, L.J., Hong, H.G. and Mallouk, T.E. 1988 *J. Am. Chem. Soc.* **110** 618–20 by kind permission of the authors and the American Chemical Society.)

the fact that a zirconium ion needs two phosphonate groups to react with and thus there will always be some unreacted phosphonate groups buried in the structure, and a gradual deterioration in the quality of the film with increasing thickness is thus likely. As it is now over three years since the publication of these two papers [392, 393] and no further work making use of this reaction has appeared, it seems likely that these problems have not been surmounted.

Tillman *et al.* [394] used a variant of the Sagiv [17] technique, in which the upper end of the OTS-like compound is terminated by the methyl ester group. The resultant layers were then reacted with $LiAlH_4$ in THF, leading to the production of terminal OH groups. The process is illustrated in Figure 6.9. These authors were able to obtain multilayers of reasonable quality consisting of up to 25 monolayers and thus, at the time of writing, hold the record for the number of layers successively deposited by the self-assembly method. Once again, however, this work was published early in 1989 and, at the time of writing, nothing further has been heard of this technique.

Putvinski *et al.* [395] have proposed a variant of the zirconium process [392, 393] in which a succession of three different reagents are applied and it is possible to form non-centrosymmetric films. They produced an initial surface of OH groups by reacting $HS(CH_2)_{11}OH$ with a gold

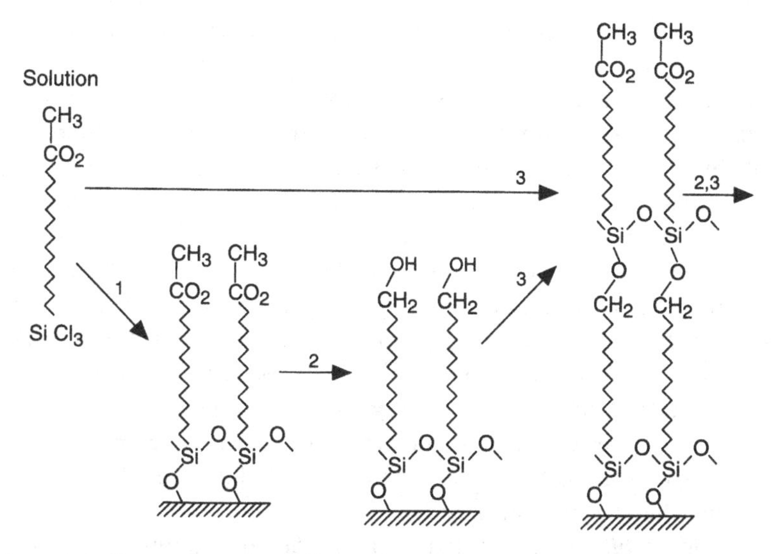

Figure 6.9. Formation of multilayers by self-assembly according to the method of Tillman *et al.* [394] The processes employed were (1) hydrolysis and chemisorption, (2) chemical activation by $LiAlH_4$ and (3) chemisorption.

surface. This surface was then phosphorylated by treatment with an acetonitrile solution of $POCl_3$ and $(CH_3CH_2)_3N$ for 6 h at room temperature. The surface was then treated with an aqueous solution of $ZrOCl_2$ and then with a solution of $HO(CH_2)_{11}PO_3H_2$, thus producing a further surface consisting of OH groups. The second, third and fourth reactions were now repeated to produce a further layer. Very substantial washing processes were required between each step, so it is doubtful whether this otherwise interesting technique is capable of much development. Furthermore, the thickness versus 'number of layers' plots obtained by this method are inferior to those obtained by the same authors using the method described in [392, 393].

Jin and Johnstone [396] have originated a method of forming multilayers by self-assembly which makes use of the fact that isocyanates react with alcohols to produce a very high yield. The initial step makes use of one of the various methods discussed above to form a substrate surface covered with OH groups. The groups on this surface can be viewed as the alcohols in the reaction shown below and will be denoted by ROH.

$$CH_2{=}CH(CH_2)_9{-}N{=}C{=}O + ROH \rightarrow CH_2{=}CH(CH_2)_9{-}NH{-}CO_2R$$

The vinyl groups now exposed on the surface can be converted to OH groups by treatment with a solution of diborane in tetrahydrofuran followed with treatment by an alkaline solution of H_2O_2. One of the other means of bringing about this process which have been mentioned above could also presumably be used. The whole process can now be repeated. These authors used water contact angles to monitor their results. The initial layer produced a contact angle of 105° and the subsequent three layers, contact angles of between 78° and 80°. They argue that, if these figures represented a deterioration of layers with the number of layers deposited, the contact angle would continue to change with each successive layer and, indeed, this appears to be a convincing argument. They also point out that the hydrogen bonding which can be expected between the oligomers formed by the process will add to the stability of the layer. In the opinion of the present writer, this process seems to be the most promising way of forming multilayers by self-assembly. It is a pity that these authors did not persevere and produce layers thick enough so that other diagnostic techniques could be used to investigate the structure obtained.

Tredgold *et al.* [397] adopted a rather different approach to self-assembly. In Equation (1.11) in Section 1.3 it was shown that, when London forces predominate, materials having different high frequency permittivities, and hence different refractive indices, will tend to separate.

This tendency is driven by the internal energy term in the free energy and opposed by the entropy term. For large fairly rigid molecules, the former term is more likely to predominate. Banks [398] has collected together the refractive indices of various straight chain perfluorinated compounds and finds figures of around 1.25. This figure can be compared with a value of 1.37 for analogous ordinary hydrocarbons. The difference between these two figures is large enough so that perfluorinated and ordinary hydrocarbons tend to separate, a fact that is well known amongst organic chemists. Now consider the two materials shown in Figure 6.10. It is relatively simple to adsorb the perfluorinated acid to a hydrophilic substrate. Hexadecane is employed as the solvent, care being taken to remove residual traces of water and other polar impurities. One now has a surface consisting of perfluorinated material which readily sheds traces of hexadecane when removed from this solvent. The substrate is now dipped into an aqueous solution of the quaternary ammonium salt and a monolayer of the latter is adsorbed in about 20 s, the two hydrophobic perfluorinated materials being attracted to one another. The substrate is now removed from the solution and carefully dried and the process is repeated. The multilayers so formed were monitored by using aluminium substrate and depositing top electrodes of gold by evaporation *in vacuo*. The capacity of the capacitors so formed should be inversely proportional to the number of layers

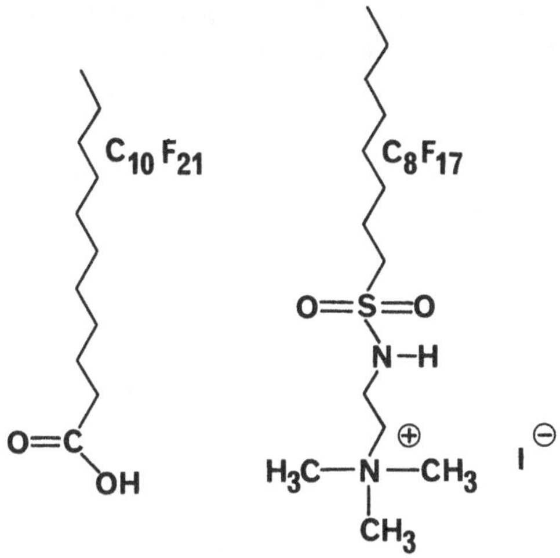

$C_{10}F_{21}$ — O=C—OH

C_8F_{17} — O=S=O — N—H — $H_3C—N^{\oplus}(CH_3)—CH_3$ $I^{\ominus}$

Figure 6.10. The perfluorinated compounds employed to form multilayers by physisorption according to the method described in [397]

deposited. Thus plotting the inverse of the capacity versus the number of bilayers should yield a straight line. Such results were obtained for the first seven bilayers but attempts to obtain satisfactory deposition for thicker systems failed. The simple theory put forward in the first part of this section predicts an exponential decay of layer quality with the number of layers deposited. In this case, however, we experience good layer quality for the first seven bilayers and then a rapid catastrophic decay after that. This is not, however, surprising as, in this case, molecules will be attracted and stabilised by the lateral action of London forces and thus a sudden decay at some critical coverage is to be expected.

Several other methods of growing oligomers in a vertical direction from a substrate have been given in the literature and are best described at this point. Thus Kubono *et al.* [399] used a vacuum coating unit to evaporate 1,10-diaminodecane at a constant speed while sebacoyl dichloride was supplied at regular intervals from a separate vacuum system. A substrate previously coated with the diamine was employed and a polyamide formed spontaneously. Initial low angle X-ray diffraction experiments indicated the presence of a regular layer structure and the regularity was much improved by annealing for 10 min at 90 °C. Ogawa *et al.* [400, 401] have used a cycle which involves initial adsorption of an OTS-like material having a terminal vinyl group and subsequent bombardment of the surface by electrons in an oxygen atmosphere under reduced pressure, which process produces a terminal OH group. Ogewa *et al.* [400, 401] used infrared spectroscopy as their principal diagnostic technique and reported producing up to five good layers.

7
Liquid crystals

7.1 Introduction

As was pointed out in Chapter 1 liquid crystals (or mesophases as they are often called) were first discovered by Reinitzer [19] in 1888 and the first proper classification of liquid crystals was made by Friedel [20] in 1922. Since that time various new categories of liquid crystals have been discovered and named. It would be impossible to give an extensive treatment of this important and wide ranging subject here but, as so many of the systems discussed in this book have a liquid crystalline structure, at least a brief treatment of the topic is essential. Furthermore, several methods of forming ordered thin organic films not treated in other chapters depend on the initial formation of a mesophase. It has been suggested that something like 10% of fine organic materials listed in a typical catalogue of such products are capable of existing in a mesophase within some appropriate temperature range or, in the case of lyotropic liquid crystals, when dissolved at an appropriate concentration in some solvent. It is thus obvious that the subject has immense ramifications and could not be pursued in any great breadth here.

Liquid crystals can initially be divided into thermotropic and lyotropic materials. The first category involves a single molecular species and exists in a temperature range which lies between the melting point of the solid phase and the temperature at which a true liquid is arrived at. In many cases this temperature range is subdivided by further phase changes into various different mesophases having distinctive properties. Lyotropic liquid crystals consist of a concentrated solution of a large molecular species in a solvent consisting of small molecules. In most of those systems which have received extensive study, the solvent is water, but many other possibilities exist. Clearly, in this case, phase changes depend on both the concentration of the solution and on the temperature. Irrespective of whether one is dealing with thermotropic or lyotropic liquid crystals, it is also possible to categorise these materials in terms of structure and symmetry.

A simple liquid can very often be thought of as an assembly of spherical molecules interacting via central forces. The nearest real approximation to this ideal system is a liquid rare gas such as argon but, even here, many-body forces occur, as was discussed in Section 1.3. However, such a picture can be helpful in considering many other simple liquids. In order for a mesophase to exist, one requires interactions which depart markedly from isotropic central forces. At this same level of simplification, one can think of molecules being represented as either prolate or oblate spheroids. The first category consist in practice of rod-like molecules, usually having a central conjugated region such as a biphenyl or stilbene group and straight hydrocarbon end groups. Such materials can form a nematic phase or one of the many smectic phases discussed below. The second category, the form of which can be approximated by oblate spheroids, consists in practice of flat disc-like molecules such as, for example, phthalocyanine derivatives and molecules of this category are capable of forming discotic mesophases. In addition to these considerations, it is possible for some chiral molecules to form liquid crystals and for the chirality to have interesting additional effects.

Let us now return to the consideration of rod-like molecules. It is intuitively obvious that, just above the melting point, molecules and their immediate neighbours will continue to point roughly in the same direction, thus producing a short range order over and above that which appears in ordinary liquids. What is less obvious is that it is possible to produce a fluid phase in which there is a long range correlation in the directionality of the molecular rods. Such a structure is known as a nematic liquid crystal providing that this directional correlation is not accompanied by other additional degrees of order. The average direction in which the long axes of the molecules point is described by a vector known as the director. In many real cases there exist other phases which lie in the temperature range between the solid and the nematic phase. In these phases the rod-like molecules not only tend to point in the same direction but also to exist in planes in which the rods lie side by side and are roughly normal to these planes. Such planar structures are known as smectic liquid crystals. Many different types of order and packing are possible within this general picture and these different structures are traditionally denoted by various capital letters. The smectic-A phase is the least ordered smectic structure whereas, by the time one arrives at the smectic-E phase, it is not easy to distinguish between a liquid crystal and a true solid. Unfortunately, as the various phases were labelled as they were recognised, it is not possible to relate degree of order to position in the alphabet in a simple manner. Figure 7.1 gives a synopsis of the various different smectic mesophases in schematic form. To understand this scheme one has to take three further factors into account.

1. It is possible for the long axes of the molecules to have a uniform tilt over macroscopic distances, as is also true in the cases of Langmuir–Blodgett films and films at the air/water interface.

2. The rod-like molecules which we have heretofore represented as prolate spheroids are better represented as ellipsoids having one very long axis and two unequal short axes. The longer of these latter two corresponds to the plane of the ring structures which occur in the central region of the molecules. It is traditional to refer to this as the short axis

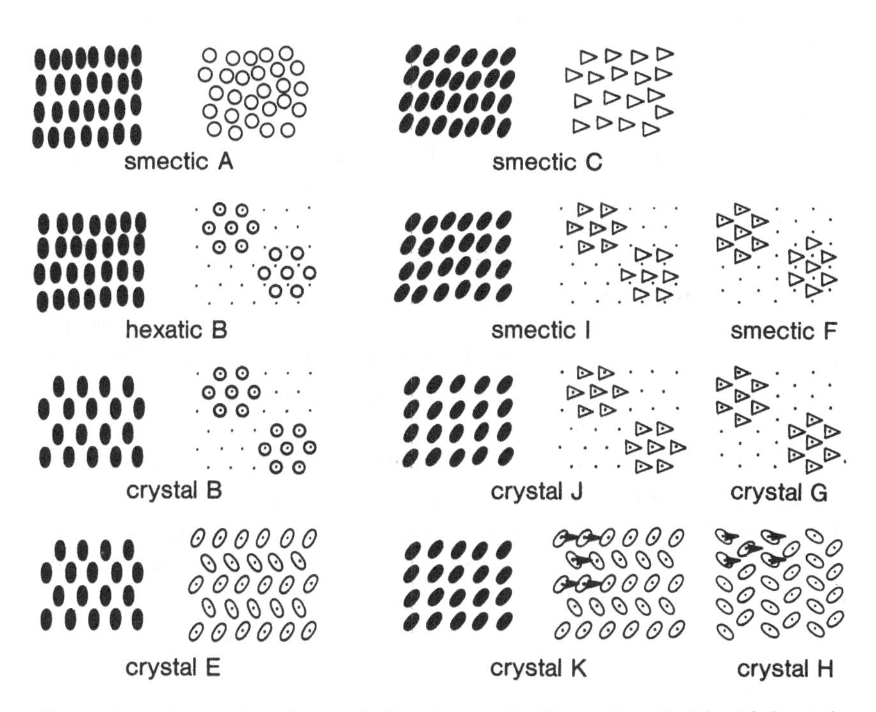

Figure 7.1. Categories of smectic liquid crystals. From *Smectic Liquid Crystals, Textures and Structures*, Gray, G.W. and Goodby, J.W.G. 1984 (Leonard Hill, Glasgow). (Reproduced by kind permission of the authors and publishers.) Side elevation and plan representation of molecular ordering in each of the smectic modifications. Triangles or arrows are used to represent tilt direction.

of the molecule. In some smectic phases the molecules are free to rotate about the long axis, but in more highly ordered systems the short axes of the molecules are arranged in some definite structure such as a herring-bone pattern.

3. The regions at the opposite ends of the long axis are usually different in some way. Sometimes this difference is rather slight but in other cases, as for example when the material consists of amphiphilic molecules, the differences are considerable. Thus, in the first case, the planes of molecules contain a random mixture of individual molecules some pointing in one direction and some in the opposite, whereas in the latter case all the molecules in a given plane point in one direction but those in alternate planes point in opposite directions, as is the case for LB films formed in the Y mode. It is now usual to distinguish the two cases by the suffixes 1 and 2 respectively. The sub-case of 2 when there is considerable interdigitation is sometimes denoted by the suffix d.

These remarks should allow the reader to understand Figure 7.1. The additional complication of chirality will be returned to later. The smectic phases are discussed in more depth by Gray and Goodby [402] and Leadbetter [403].

Turning now to those molecules whose shape can be approximated by oblate spheroids, one arrives at the discotic phase. Here the average of the normals to planes of the molecules corresponds to the director. A fluid phase in which these normals point in roughly the same direction over a macroscopic distance is said to be discotic. If this factor is the only degree of order, the material is said to be in the nematic discotic phase. If, in addition, the discs stack in regular columns, the material is said to be in the columnar discotic phase. Such structures have been discussed in Section 4.5.1.

In order to avoid this section becoming too abstract, a selection of molecules which can form nematic or smectic liquid crystals is illustrated in Figure 7.2. For a discussion of how particular molecular structures lead to formation of particular mesophases, reference should be made to the work by Gray and Goodby [402] already cited or to Chapters 1 and 12 of *Molecular Physics of Liquid Crystals* edited by Luckhurst and Gray [28].

7.2 Liquid crystals: theoretical considerations

Liquid crystals have attracted considerable theoretical study. Much of this has been of a secondary nature and has explored the properties of

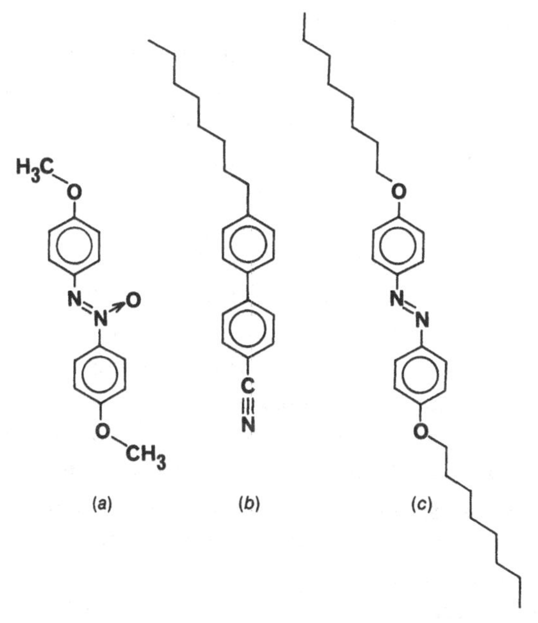

Figure 7.2. Typical molecules which form mesophases at room temperature: (*a*) is in the nematic phase, (*b*) is in the smectic-A phase and (*c*) is in the smectic-C phase.

these materials given that the materials already exist. Of more interest in the present context are those studies which have made an attempt to understand how these ordered structures are brought about in the first place and how the phase changes which are associated with them take place. Early studies made use of the molecular field approximation. This approximation was first introduced by Weiss in 1907 [404] and has been applied to almost every order/disorder phenomenon which has been studied since. The procedure involved in using this approximation is to concentrate attention on one typical particle or molecule. The supposed potential in which this molecule moves is then defined in terms of one or more parameters. The probability distribution functions of the position and orientation of the molecule are then calculated using conventional statistical mechanics. It is now assumed that the neighbouring molecules, interaction with which brings about the potential which was originally assumed, also obey this distribution function and the parameters defined above are adjusted so that the various assumptions are made self-consistent. It is evident that such a procedure will go some way towards describing a cooperative system but that it leaves much to be desired. To see why this should be so, it is helpful to consider the application of the molecular field approximation to the original problem to which it was applied. Suppose one considers its application to the Ising model for a ferromagnet in which a single spin resides at each point on a regular lattice and that this spin has two possible states, namely up or down. There is an interaction energy between neighbouring spins which depends on whether they are parallel or antiparallel. The potential in which one chosen spin exists depends on the relative probability of the neighbour spins being up or down and this probability in turn depends on the probability of the chosen spin being up or down. Such a treatment of this system has the weakness that it ignores the fact that, if the chosen spin is in the up position, the neighbours will have a greater probability of being in the up position and *vice versa*. Interactions of this kind between the neighbours surrounding the chosen central molecule are also ignored. In 1944 Onsager [405] gave an exact statistical mechanical treatment of a two-dimensional Ising lattice and subsequently much progress has been made in treatments of the three-dimensional problem. These more sophisticated treatments show that the molecular field method gives a reasonable prediction of the behaviour of the Ising model at temperatures far from a transition point but gives a rather poor description of what happens near such phase transitions.

Early theoretical treatments of liquid crystals were not surprisingly based on the molecular field approximation. However, it is neccessary to make assumptions about the pair potential employed in the calculation and it is impossible to know whether the predictions of a particular model really arise from the pair potential employed or whether they arise, at least in part, from the deficiencies of the basic approximation employed. The general problem is so complex that a better *mathematical* treatment of the molecular interactions in a liquid crystal is out of the question. However, with the introduction of ever more powerful computers, it has become possible to carry out meaningful numerical simulations of model liquid crystals.

7.3 Computer simulations

Zannoni [406] has given a very useful discussion of the various techniques involved in computer simulation of liquid crystals. However, since that work was published, far more powerful computers have become available and problems have been tackled with success which would have been impracticable to tackle a few years ago. Recently Allen and Wilson

[407] have published a review of the results of calculations made using more modern powerful computers and some of these results will be given below. A brief discussion of the basic methods employed in computer simulation of liquid crystals is of interest as such techniques have also been applied to models of films at the air/water interface and to models of self-assembled films. The results obtained in two particular cases of the simulation of LB films will be discussed at the end of this section.

There are two basic approaches to the computer simulation of liquid crystals, the Monte Carlo method and the method known as molecular dynamics. We will first discuss the basis of the Monte Carlo method. As is the case with both these methods, a small number (of the order hundreds) of molecules is considered and the difficulties introduced by this restriction are, at least in part, removed by the use of artful boundary conditions which will be discussed below. This relatively small assembly of molecules is treated by a method based on the canonical partition function approach. That is to say, the energy which appears in the Boltzman factor is the total energy of the assembly and such factors are assumed summed over an ensemble of assemblies. The summation ranges over all the coordinates and momenta which describe the assemblies. As a classical approach is taken to the problem, the summation is replaced by an integration over all these coordinates though, in the final computation, a return to a summation has to be made. If one wishes to find the probable value of some particular physical quantity, A, which is a function of the coordinates just referred to, then statistical mechanics teaches that this quantity is given by

$$\langle A \rangle = \frac{\int e^{-E/kT} A \, d\tau}{\int e^{-E/kT} \, d\tau} \tag{7.1}$$

where E is the total energy of an assembly. E is a function of all the coordinates and conjugate momenta and $d\tau$ represents integration over all these coordinates and momenta with limits imposed by the 'box' in which an assembly is contained. If A is a function of the coordinates which define the potential energy only and not of the conjugate momenta, it is possible to factor out the parts of both numerator and denominator in (7.1) which depend on momenta only and arrive at

$$\langle A \rangle = \frac{1}{Z} \int e^{-U/kT} A \, d\tau \tag{7.2}$$

where U is the potential energy and Z is given by

$$Z = \int e^{-U/kT}\, d\tau \tag{7.3}$$

Here one arrives at expressions which are equivalent to those in the initial macrocanonical approach but with the expression for the total energy replaced by the potential energy. One can now apply the principle of detailed balancing to these expressions. Thus, if the probability of the potential energy changing from U_i to U_j is given by P_{ij} and the converse probability by P_{ji}, then these expressions will be related by the equation

$$\frac{P_{ij}}{P_{ji}} = \exp\,[-(U_j - U_i)/kT] \tag{7.4}$$

These statements follow directly from Maxwell–Boltzman statistics. The Monte Carlo method now takes the probability of all downward transitions as unity so that, if P_{ij} represents the probability of an upward transition, it is given by

$$P_{ij} = \exp\,[-(U_j - U_i)/kT] \tag{7.5}$$

The following approach now seeks to find an equilibrium assembly. Arbitrary and random trial changes of the coordinates of one particle at a time are made and then the resulting configuration is 'accepted' or 'rejected' according to a procedure explained below. If this process is carried on for a long enough time, one eventually arrives at an approximation to thermodynamic equilibrium and the value of A corresponding to the value which would be given by (7.2) can be calculated. As was pointed out above, the probability of a downward transition is always taken as unity. To decide if the upward transition P_{ij} in (7.5) is to be accepted or rejected, the quantity on the right hand side of (7.5) is compared with a randomly generated number between 0 and 1 and if it is greater than this random number the transition is accepted and otherwise it is rejected.

After running a programme to carry out the processes outlined above for a sufficiently long time, an approximation to the equilibrium state is arrived at and the quantity $\langle A \rangle$ may be calculated.

When a model which makes use of a hard core potential is used, the above procedure can obviously not be employed. In that case, random changes of the configuration are tried and are only accepted if they do not involve an overlap of the volumes occupied by two different molecules, as such an overlap would clearly imply an infinite value of the potential energy.

To overcome the limitations imposed by the small number of particles and the small size of the 'box', cyclic boundary conditions are employed. The idea of cyclic boundary conditions is rather obvious in a one-dimensional system as the one-dimensional 'box' is simply imagined to be bent back on itself. As one is interested in the interaction between pairs of particles, it is necessary to avoid multiple interactions and the convention is adopted that a given particle is assumed to interact with the nearest version of another particle. To extend this idea to a three-dimensional system, one imagines a structure made from $3 \times 3 \times 3$ boxes. The central box is taken as the starting point and particles in this box are taken to interact with either particles in this box or particles in one of the 26 other boxes, according to rules which are a straightforward extension of the one-dimensional cyclic process.

The molecular dynamics method uses similar boundary conditions and employs one of the following approximations to solve the problem of an assembly of molecules contained in a 'box' and whose motions are governed by the laws of classical dynamics.

In the case that the energies of interaction between molecules are continuous functions of the coordinates (the so-called soft potential), the equations of motion of each individual molecule are assumed, during the time period t to $t+\Delta t$, to be governed by forces calculated from the potential function $U(t)$. The differential equations of motion are approximated to finite difference equations and the finite changes in each coordinate and conjugate momentum corresponding to an increment of time, Δt, are calculated. This process is repeated many times and, for sufficiently small Δt provides a good approximation to the solution of this many-body problem. If the process is carried out for a long enough time, a distribution of coordinates and momenta is arrived at which corresponds approximately to thermodynamic equilibrium. Though the process described above indicates the general approach actually adopted, the difference equations usually used are more sophisticated and were originally proposed by Verlet [408].

The values of the coordinates and velocity of the centre of mass of a molecule, r and v at the time $t+\Delta t$ are expanded as a Taylor series through terms in $(\Delta t)^2$. Making use of these expressions for both $+\Delta t$ and $-\Delta t$ and Newton's equation of motion, simple manipulation yields the Verlet algorithm

$$r(t+\Delta t)+r(t-\Delta t)=2r(t)+\frac{(\Delta t)^2F}{m} \quad (7.6)$$

$$v(t)=\frac{r(t+\Delta t)-r(t-\Delta t)}{2\Delta t} \quad (7.7)$$

where F is the force acting on the particle and m is the mass of the particle. Similar processes can be applied to the angular coordinates.

In the case of hard core interactions the treatment of the problem must be different. Again, finite difference equations are employed and the configuration is allowed to evolve until a collision takes place. The consequences of the collision are predicted according to the laws of classical mechanics and, again, the configuration is allowed to evolve until the next collision takes place.

We turn now to the results of computer simulation. Early work in this field was based on the lattice model in which each molecule is assumed to be located at a point of a three-dimensional lattice and the variables are the orientational coordinates. Until relatively recently the computers available could not tackle a more realistic model. However, the lattice model does, to some extent, assume the answer to the problem which one is examining and is unable to make predictions about the interesting questions concerning the formation and stability of particular mesophases.

In the hard core approach, an assembly of either spheroids or spherocylinders (a spherocylinder is a cylinder capped at either end by a hemisphere) is assumed to be enclosed in a 'box'. The pressure exerted by the 'box' takes the place of the attractive potential which has been ignored and makes it possible to achieve condensed phases. The problems of assemblies of both oblate and prolate spheroids have been studied by Frenkel *et al.* [409] using the Monte Carlo approach and by Allen *et al.* [410] using molecular dynamics. As the volume available to the molecules is reduced, a point is reached at which a nematic phase becomes stable. The density at which this phase change takes place depends on the degree of eccentricity, ϵ, of the spheroids but, for both oblate and prolate spheroids, a nematic phase is possible. For a given density, the values of ϵ and $1/\epsilon$ at which the phase changes take place for the two systems respectively are, at least approximately, equal. The phase change is first order. No smectic phase appears to exist.

Hard spherocylinders behave differently. Frenkel *et al.* [411] have studied systems consisting of these objects having length to diameter ratios of both 3 and 5 and have employed both the Monte Carlo method and molecular dynamics. They find a more realistic behaviour with an isotropic phase at low densities, a nematic phase at higher densities, a smectic phase at still higher densities and a true solid at a density corresponding to 0.7 of the density which is arrived at when the system is closely packed. (The results quoted are for a length to diameter ratio of 3.)

Frenkel [412] also examined the behaviour of systems consisting of hard spheres in which the top and bottom have been cut off to form disc-like structures. Here it is possible to demonstrate the existence of both discotic nematic and hexagonal columnar discotic phases.

It is truly remarkable that these rather simple models are able to predict these relatively complex phases. The existence of a nematic phase appears to be likely from a purely intuitive approach but the existence of a smectic phase, at least for such a simple model, is by no means intuitive. These results have important implications for the understanding of other layer structures such as LB films.

Soft non-spherical potentials are one step towards a more realistic model. Luckhurst *et al.* [413] have used a potential having the prolate spheroidal symmetry discussed above but which is based on the well known Lennard-Jones or twelve–six potential. This involves an attractive $1/r^6$ potential based on London forces and a repulsive $1/r^{12}$ potential. Once again it is possible to predict the existence of a smectic phase.

Recently attempts have been made to study systems using yet more realistic potentials and it is likely that considerable progress will be made in this direction in the near future.

The analogous problem of simulating Langmuir and LB films has recently received some attention but is more complex than the treatment of liquid crystals as one is dealing with an interface and, in the case of Langmuir films, with two different kinds of molecule. Bareman and Klein [414] and Moller *et al.* [415] gave treatments of LB monolayers, both of which make use of the molecular dynamics approach. In the first of these studies, an assembly of 90 molecules was examined and in the later study the number was 64. Both papers predicted a phase change from a tilted to a vertical configuration as the area per molecule is decreased as, of course, is observed experimentally. This change in structure is, according to both groups of authors, driven by local interatomic interactions and is not predicted by a model in which the molecules are represented as rods. However, as has been shown in earlier chapters in this book, a wide variety of rod-like molecules tend to pack in tilted structures even when the rods do not consist of simple hydrocarbon chains. Thus it is at least possible that some more general mechanism governs rod tilt.

7.4 Chirality

The concept of chirality is a commonplace to chemists but may not be so to other readers of this book. Before discussing the implications of chirality in liquid crystals, a brief definition of this concept will accordingly be given. It is not possible to bring a chiral object into coincidence with its mirror image by any combination of simple translations and rotations. A familiar example is a helix as exemplified by a screw thread. No amount of twiddling will turn a right hand screw into a left hand screw. A condensed phase consisting of chiral molecules will thus have a helical structure though the pitch of the helix may be large relative to molecular dimensions. Chiral liquid crystal phases can be created by the use of suitable chiral molecules or by mixing these with molecules which would lead to an ordinary nematic phase if used alone. It is also possible to form smectic phases with chiral molecules but a discussion of such systems will be deferred until the effect of chiral molecules on the nematic phase has been discussed.

The most interesting characteristic of chiral nematics is their interaction with light. A full and rigorous account of this topic is beyond the scope of this book. However, consider a linearly polarised light wave propagating along the axis of a helical liquid crystal phase. The direction of the director of the liquid crystal will rotate through an angle of 360° over a distance corresponding to the pitch of the helix and the index of refraction will depend on whether the electric vector is along or at right angles to the director. Over a sufficient distance, the direction of the electric vector will rotate, as is the case in all chiral media. However, over a short distance, one can ignore such rotation in this simple approximate treatment.

Now consider the case when the pitch and the wavelength of light in the medium are equal. The direction of the electric vector will come into coincidence with the director once for every half wavelength of propagation. The rotation of this direction of polarisation determines the effective refractive index and thus there are two sinusoidal periodic variations of the refractive index for every wavelength travelled. The result of this variation is to provide a grating which diffracts the light. For moderate angles of propagation relative to the helix axis, this statement is still approximately true. Simplistically one might now apply the Bragg diffraction condition, $n\lambda = 2d \sin\theta$. However, as there are two periods of change of effective refractive index for one period of pitch, $2d = p$, where p is the pitch. Also, as the variation of refractive index is sinusoidal, only the first Bragg order will appear, by analogy with the case treated in Section 2.3 and Equation [2.11]. Thus one arrives at the expression

$$\lambda = p \sin\theta \tag{7.8}$$

As the pitch of many cholesteric liquid crystals is a function of temperature, this phenomenon has been exploited to make thermometers.

Chiral molecules which form smectic liquid crystals are often capable of forming structures in which the electric dipoles associated with the molecules all point approximately in the same direction in a particular region but in which this direction rotates as one moves in a direction normal to the smectic planes. Such materials are rather misleadingly referred to as ferroelectric liquid crystals. The mechanism responsible for this effect is illustrated in Figure 7.3. The molecules tilt into a smectic-C phase due to their structure as illustrated. Dipoles associated with the molecules are supposed to point in a direction normal to the page. Thus, if the molecules all have the same handedness the dipoles all point in he same direction. This description is an oversimplification as the molecules rotate about their long axes but point preferentially in the manner indicated. This phenomenon has been successfully applied to

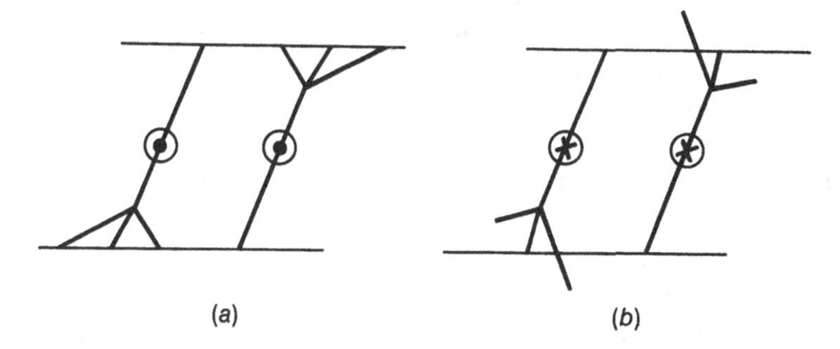

Figure 7.3. Schematic diagram to illustrate the mechanism of ferroelectric liquid crystals. It is supposed that the molecules have a structure which encourages a tilt as in (*a*). It is further supposed that (in this simple illustration) there exists an electric dipole at right angles to the molecular axis which points out of the paper in (*a*), $\odot$, and into the paper in (*b*), $\otimes$. It is supposed that the configuration shown in (*a*) represents the lower energy state.

the construction of optical display devices and this topic has been reviewed, for example, by Coates [416].

7.5 The application of mesophases to the formation of ordered thin films

Some of the phenomena discussed in earlier sections of this book involve forms of matter which might well be described as smectic liquid crystals but are better dealt with under other headings. However, there are three particular processes which do not fit into these other categories but which certainly make use of liquid crystalline behaviour and will thus be examined here.

Lyotropic rigid rod polymers can behave as nematic liquid crystals. In particular poly(γ-benzyl-L-glutamate) (PBLG) has been studied by various workers in this context. Tredgold and Ali-Adib [417] studied the formation of thin solid films of this material by casting a thin layer of a nematic solution of it in a solvent which could subsequently be polymerised by irradiation by ultraviolet light (3-chloro-2-chloromethyl-1-propene).The polymer chains were aligned by making use of the diamagnetic properties of the benzyl groups and the application of a 5 T magnetic field. The difficulty of using PBLG for this experiment is that this polymer is chiral and thus the individual polymer chains tend to twist with respect to one another which prevents perfect alignment. Subsequently Ali-Adib *et al.* [418] performed similar experiments using a nematic solution of a butoxy substituted phthalocyaninato-polystiloxane in a solvent consisting of 1,1,1,3,3,3-hexafluoro-2-propanol. The structure of this type of polymer is shown in Figure 5.15 and LB films made from them are discussed in Section 5.5. In this case evaporation of the solvent takes place at a controlled rate and the final solid film thus consists entirely of the polymer. A dichroic ratio of 7.3 was obtained which compares favourably with the value of 2.3 obtained by Sauer *et al.* [285] using the LB method. Films having a thickness of 50 nm were readily formed by this technique.

The diamagnetic forces exerted on a single phthalocyanine ring by a field of 5 T are too small to have an effect large enough to compete with thermal disturbances. However, the use of polymers and the cooperative effect involved in a nematic liquid crystal totally alter the situation. The substrate is held in a horizontal position and the magnetic field is directed so as to be tangential to the substrate. The position of minimum energy for a plane conjugated structure in a magnetic field is such that the direction of the field is tangential to the plane. Thus the final aligned structure is such that the polymer axes are at right angles to the field and lie in a horizontal direction. The outer layer of the polymer consists of the butyl groups which are, of course, hydrophobic. If the substrate is rendered hydrophobic, the film adheres well to it and a good clear film is obtained. If the substrate is hydrophilic, the film adheres well to itself but poorly to the substrate and, as the solvent evaporates, fine cracks appear in a direction parallel to the polymer chains.

Amador *et al.* [419] studied freely suspended films of liquid crystals by diffraction of synchrotron X-ray radiation. They formed five molecular layers of N-[4′-(n-heptyl)benzylidene]-4-(n-heptyl)aniline and found the following behaviours.

At 78.14°C all five layers are in the smectic-C phase.
At 77.67°C the two surface monolayers are in the smectic-I phase and the three interior layers are in the smectic-C phase.
At 69.09°C the entire film is in the smectic-I phase.

These authors also quote earlier work devoted to the study of freely suspended smectic liquid crystals.

Decher *et al.* [420] have developed the idea of using freely suspended smectic liquid crystals and have used it to form thin films on solid substrates. A thermotropic smectic liquid crystal is drawn across an aperture in a solid support and is capable of bridging the aperture (which can be up to 15 mm in diameter) with a film which can be between two and several hundred layers thick. The film consists of a smectic structure with the layers lying in the plane of the film. The film is formed a short distance above a solid substrate and the apparatus is constructed so that a difference of pressure between the two sides of the film can be used to force the film down in contact with the substrate. These authors have thus formed good quality films up to an area of about 1 cm^2. In the work described the material used was ethyl-4′-n-octyloxybiphenyl-4-carboxylate.

Agarwal *et al.* [421, 422] showed that it is possible to form Y layers of stearic acid by evaporation *in vacuo* on to a suitable substrate. These authors concluded that such multilayers are inferior in quality to those produced by the LB technique. However, Jones *et al.* [145] found that slow evaporation and careful control of the substrate temperature make it possible to produce high quality films (as proved by X-ray diffraction) by this technique. They also succeeded in producing good Y layers of short perfluorinated carboxylic acids ($C_9F_{19}COOH$ and $C_{11}F_{23}COOH$).

These materials are too short to form stable layers at the air/water interface and thus have never been deposited by the LB technique. The best results were obtained with a substrate temperature of between −10 and −30 °C and the quality of the films was sufficiently good that Bragg peaks of up to 14th order were observed. Even order peaks predominated, indicating that the important periodicity corresponded to the thickness of a monolayer, as would be expected from the high electron densities of the perfluorinated chains. The films were now immersed in 10^{-2} M $AgNO_3$ for 20 min and Ag^+ ions diffused into the hydrophilic layers associated with the carboxylic acid groups. This is a technique adapted from the work of Barraud *et al.* [423]. As would be expected, the predominant Bragg peaks were now of odd order, indicating that a true Y layer structure existed. Good Y layers of the azobenzene materials shown in Figure 4.4(*a*) were also produced by this method and the X-ray diffraction results obtained from them corresponded very closely to those obtained from film formed from the LB technique.

For good Y layers to be formed by evaporation *in vacuo* one requires conditions such that the bulk material is stable but the surface region is sufficiently liquid that rearrangement of the molecules into a layer structure is possible. It thus seems likely that, in the temperature range of interest, the surface region behaves as a smectic liquid crystal. It is not yet clear how far the bulk regions of such multilayers should be thought of as liquid crystal-like in nature.

Many other organic materials have been deposited by evaporation in vacuo but usually form either a polycrystalline or an amorphous structure. However, Hoshi *et al.* [424] have made some progress in depositing epitaxial films of lutetium diphthalocyanine on to single crystals of potassium bromide. Here again the temperature of the substrate is critical but only relatively small areas of continuous crystal have been obtained.

8
Biomolecules

8.1 Introduction

The study of order in structures involving biomolecules divides naturally into two parts. On the one hand, one can consider ordered structures *in vivo*, and on the other hand in man-made systems. The obvious example of thin organic films in the former category is the cell plasma membrane (the term for the exterior membrane of a cell). In 1925 Gorter and Grendel [425] suggested that the cell membrane consisted of a bilayer of lipid molecules with the hydrophilic ends facing outward and the hydrophobic ends facing one another in the interior of the membrane. (The structures of some common lipids are shown in Figure 8.1.) It was a long time before this postulate was definitely confirmed but it is now generally accepted that the plasma membrane is roughly of this general form. The main modifications of this picture are as follows.

(a) Many membrane-bound enzymes penetrate the plasma membrane and are stabilised by the fact that the surface of the enzyme consists of two hydrophilic end regions and an intermediate hydrophobic region. These enzymes take part in the transport of particular substances across the membrane and in various important processes in which energy is stored or interchanged by the medium of ion transport. This topic is returned to below.

(b) A number of glycoproteins are incorporated into the membrane and are involved in cell recognition processes.

(c) There exists a network of proteins both on the surface of the membrane and in the interior of the cell which provides mechanical strength and stability and which is known as the cytoskeleton.

The functions of materials in these three groups may overlap.

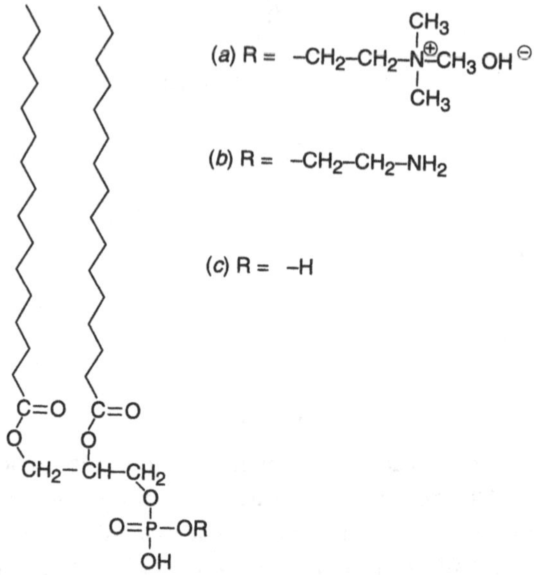

(*a*) R = $-CH_2-CH_2-N^{\oplus}(CH_3)_3\ OH^{\ominus}$

(*b*) R = $-CH_2-CH_2-NH_2$

(*c*) R = $-H$

C=O C=O
O O
$CH_2-CH-CH_2$
O
$O=P-OR$
OH

Figure 8.1. The structure of some common phospholipids: (*a*) phosphatidylcholine, (*b*) phosphatidylethanolamine and (*c*) phosphatidic acid.

The lipid part of the membrane is essentially a two-dimensional liquid in which the other materials are immersed and to which the cytoskeleton is anchored. This last statement is not totally correct, as some membrane bound enzymes require the proximity of particular lipids to function properly and are thus closely bound to them. Simple bilayers formed from lipids in which both hydrocarbon chains are fully saturated can have a highly ordered structure, but for this reason tend to be rigid rather than fluid at physiological temperatures. Natural selection has produced membranes which consist of a mixture of different lipids together with other amphiphilic molecules such as cholesterol and some carboxylic acids. Furthermore, in many naturally occurring lipids, one hydrocarbon chain contains a double bond and is thus kinked. Membranes formed from a mixture of such materials can retain a fluid structure. The temperature at which such membranes operate determines a suitable mixture of lipids so that a fluid but stable structure results at this temperature. It will be seen that the lipid part of a membrane must, apart from its two-dimensional character, be *disordered* to do its job. However, the membrane bound proteins have a degree of order, as will be discussed below.

The plasma membrane totally encloses the interior of a cell but clearly many different molecules must be able to cross this barrier if the cell is to continue to live. The methods of transport can be divided into three categories.

(a) Small uncharged molecules such as water and oxygen can simply diffuse through the fluid part of the membrane.

(b) Larger uncharged molecules are assisted in their passage through the membrane by specific enzymes.

(c) The case of small ions is one of particular interest. It was shown by Born [426] that, if an ion of effective radius, r, and charge, q, is immersed in a continuous medium of dielectric constant, ϵ, the energy of interaction between the medium and the ion is given by

$$E = -\frac{q^2}{8\pi\epsilon_0 r}\left(1 - \frac{1}{\epsilon}\right) \tag{8.1}$$

Now the dielectric constant, of water at typical physiological temperates is about 80, whereas the dielectric constant of the inner hydrocarbon part of the membrane is about 2. Putting the accepted radii of small cations such as sodium or potassium into Equation (8.1), one arrives at differences of interaction energy of the ion as between water and lipid of between 3 and 4 eV. This means that, in order to move into or out of the cell, the ion needs an energy of the order of 120 kT. Even if we treat this figure as only a rough estimate (which it almost certainly is) it is clear that direct ion transport across a membrane is a most improbable process and that, for an ion, the membrane constitutes an insulating barrier. It is thus possible to build up potentials of up to about 180 meV across membranes with only a small leakage current. Special membrane bound enzymes exist which can mediate the transport of ions across membranes and which couple the energy gained or lost to particular biochemical reactions. Many of the most important interactions both in plants and in animals which couple together processes involving changes in stored energy depend on such enzymes. So does the operation of the nerve axon. These topics are treated at length in modern texts on cell biology.

8.2 Ordered arrangements of proteins in membranes *in vivo*

With the exception of the purple membrane produced by *Halobacterium halobium*, the order discussed in this section does not involve an exact regular geometrical structure. However, membrane bound enzymes and structure proteins associated with membranes are arranged in ways which are clearly defined and which involve definite and regular interactions between different kinds of molecule. Only thus can they carry out

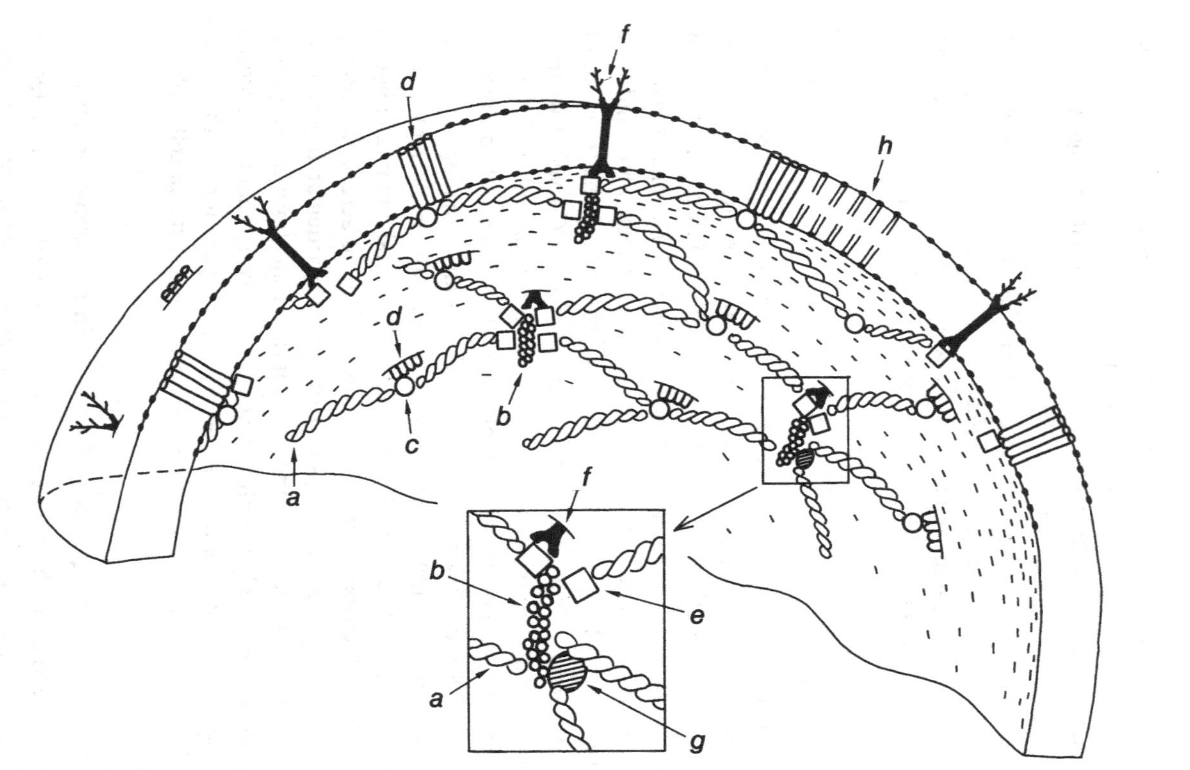

Figure 8.2. Schematic diagram of an erythrocyte membrane viewed from inside. The scale of the molecules has been expanded relative to the scale of the cell by about two orders of magnitude. *a*. Spectrin. *b*. Actin. *c*. Ankyrin. *d*. Anion transporter. *e*. Protein 4.1. *f*. Glycophorin. *g*. Adducin. *h*. Lipid bilayer.

their functions. As an example of such interactions, the membrane of the erythrocyte (red blood cell) will be discussed. This membrane has been extensively studied because a well established preparation can be used to extract the erythrocytes from fresh ox blood and then to break them open and finally to separate the membranes from their contents. The resultant free plasma membranes are known as erythrocyte ghosts. Much of the present available knowledge concerning these materials has been collected in a volume edited by Agre and Parker [427].

Figure 8.2 gives a schematic picture of a portion of an erythrocyte membrane. On the interior or cytoplasmic side of the membrane there exists a network formed from the proteins spectrin and actin. This network is bound at many points to a protein, ankyrin, which penetrates the lipid bilayer and thus serves to couple the spectrin–actin network to the lipid membrane. Each ankyrin molecule is also associated with an enzyme which assists the transport of anions across the membrane. This dual function of a protein is entirely typical of biological phenomena where a material originally evolved for one purpose finds a secondary purpose. Crick [428] has pointed out how such behaviour leads to the existence of chemical reactions and structural arrangements which are often apparently needlessly complicated. The application of Occam's razor to the understanding of biology is thus unreliable. In addition to ankyrin the glycoprotein glycophorin A also penetrates the lipid membrane and provides another anchorage for the spectrin–actin network. The outer polysaccharide end of glycophorin probably has a recognition function and thus we have another example of a dual purpose molecule.

The major network protein in the erythrocyte is spectrin. This consists of two distinct peptide chains of 260 000 and 225 000 Daltons respectively that are intertwined side-by-side to form a unit about 100 nm long. Two of these units join end to end to form a tetramer about 200 nm long, which is the principal form in which this molecule appears in the erythrocyte. Actin is a relatively short oligomer which serves to bind together spectrin molecules at various points, at each of which between five and eight spectrin molecules converge. Each of these points involves a complex in which glycophorin, a material known as protein 4.1 and several other proteins all take part. Protein 4.1 appears to be the 'glue' which helps spectrin and actin to bind together. Various other proteins have some part in this network but a discussion of their postulated functions would be out of place here. Without going into details, however, it should be stressed that the whole protein network is dynamic rather than static in nature and that binding by its various components one to another is influenced by the chemical 'climate' in the cytoplasm. Similar protein networks exist in many other kinds of cell but are less well characterised. They are often connected to an inner cytoskeleton consisting of various fibre-like structures, some of which, the micro-tubules, take part in the transport of molecules within the cell.

The general sketch of the cytoskeleton given in the last paragraph applies to eucaryotic organisms, in which the cell nucleus is contained in a separate inner membrane. All the higher organisms belong to this category, but bacteria are procaryotic and lack a nuclear membrane and have a simpler structure in other respects. They have a relatively robust *outer* cytoskeleton, the detailed nature of which depends on the particular category of bacteria.

The organism *Halobacterium halobium* has been extensively studied by Stockenius and his collaborators [429–34] and subsequently has attracted the attention of many other research workers. This bacterium and a number of related species have very strange and interesting properties. They occur in highly saline lakes and pools and require a medium having a concentration of between 3 and 5 M NaCl for growth. They also require relatively high concentrations of potassium and magnesium ions. When the surrounding medium contains adequate oxygen they can metabolise nutrients in this medium but when the oxygen concentration is low the bacteria grow patches of membrane having a diameter of up to about 1 μm which are purple in colour and have a photosynthetic function. These patches absorb light in a region centred on 560 nm and pump protons in an outward direction across the membrane, at the same time changing their absorption spectra for a period of a few milliseconds [433]. This change in absorption spectra is thought to be due to a change in conformation which is associated with the pumping process. The resultant proton gradient across the membrane is believed to drive the enzyme ATPase located elsewhere in the organism's membrane and thus produce the high energy material ATP which acts as the fuel for most cellular activities requiring energy. The presence of a proton gradient acting as a coupling agent in a chain of reactions involving the transduction of energy was first postulated by Mitchell [435] and now believed to be a fundamental process in photosynthesis and in the operation of mitochondria in both plants and animals. The process which takes place in *Halobacterium halobium* is more direct and simpler than the photosynthetic process which takes place in the higher plants and has thus attracted a great deal of study.

Our present interest in the purple membrane arises from the fact that it appears to be formed from a single species of protein which is arranged in a regular two-dimensional crystalline structure, a state of affairs which is extremely rare in living tissue. About 0.75 of the purple membrane (by weight) consists of this protein and the remainder consists of lipids. Unwin and Henderson [436] made a detailed study of the structure of this membrane using electron diffraction. Just as is the case for an X-ray structural study, the main problem to be overcome is the determination of the phases. These authors supplemented their diffraction results with Fourier analysis of electron micrographs of the material in order to overcome the phase problem. A contour map of a two-dimensional projection of electron density obtained by them is shown in Figure 8.3. Subsequently Henderson and Unwin [437] extended their technique to study tilted membranes and were thus able to obtain a reasonable three-dimensional picture of the membrane structure. Each molecule consists

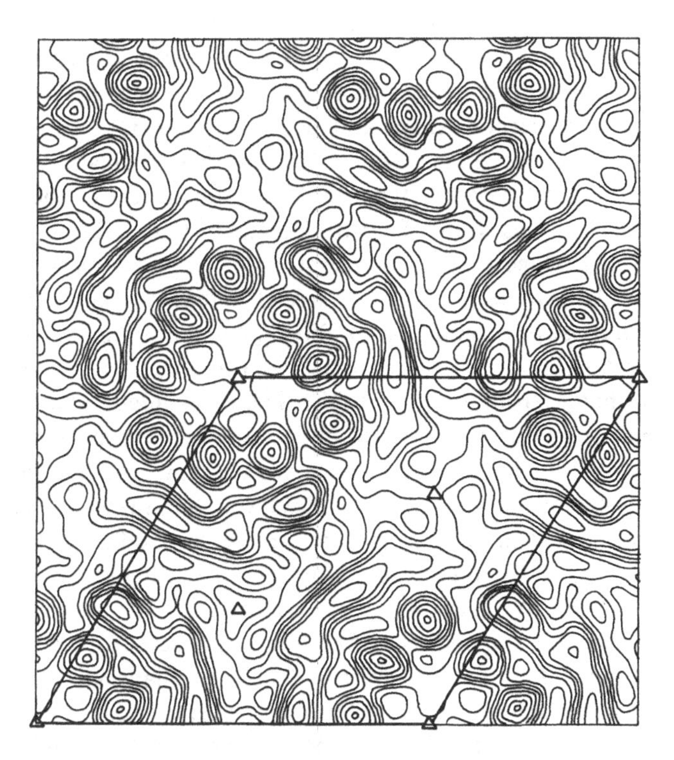

Figure 8.3. The purple membrane. Contours of constant electron density are shown. (This is a simplified version of a figure which appears in Unwin, P.T.N. and Henderson, R. 1975 *J. Mol. Biol.* **94** 425–40 and is reproduced by kind permission of the authors and Academic Press.) See the text for further discussion.

largely of seven α-helices whose axes are roughly normal to the membrane plane. The molecules are grouped in set of three. In each molecule three of the helices are almost totally vertical and thus show up particularly strongly in the contour map. These particular helices, grouped with the equivalent helices in a triad of molecules, form the sets of nine prominent objects seen in Figure 8.3. Subsequent work by Blaurock and King [438] and by Hayward *et al.* [439] confirms these results. There is an asymmetry of structure with respect to the plane of the membrane which the directionality of the proton pump would lead one to expect.

8.3 Lipids

The term lipids was originally used to describe any molecules of biological origin which are soluble in organic solvents but not in water. In the present context we will be interested mainly in the phospholipids which are diesters of phosphoric acid. Of the two compounds to react with phosphoric acid, one is often a diglyceride and the other is usually choline or ethanolamine (see Figure 8.1). The resulting material consists of two long hydrocarbon chains attached to a polar head group which is oriented roughly at right angles to the axes of these chains as illustrated in Figure 8.1. Such materials form the main constituent of the plasma membrane. Numerous studies have been made of both naturally occurring lipids and analogous synthetic materials. It is indeed surprising that these materials have been studied with such intensity under conditions remote from those pertaining *in vivo*. Silver [440] has given a useful overview of what is known of the structure of three and two-dimensional formations made from phospholipids. These include crystals, three-dimensional structures consisting of alternating bilayers and thin layers of water (known as hydrated bilayers), micelles and true bilayers. It is the latter systems together with monolayers at the air/water interface which are of most interest to us here. However, studies of hydrated bilayers have provided important information which would otherwise have been difficult to obtain. This is particularly true of the application of X-ray diffraction and neutron diffraction (using selectively deuterated lipids) to the elucidation of the electron density of bilayers as a function of the coordinate normal to the bilayer plane. Hydrated bilayers have also been useful in the investigation of thermal disorder in bilayers.

In Section 2.1 a parameter S, was defined which measures the degree to which all the chains in a multilayer lie parallel to one another and hence to the director vector which defines their average direction. A hydrated bilayer system provides a sufficient volume of parallel bilayers to allow S to be investigated by nuclear magnetic resonance. The theory of this subject is treated, for example, by Seelig [441] and will not be discussed here. It is sufficient to note that, if a hydrocarbon chain is partially or totally deuterated, then the deuterium quadrupole resonance spectrum consists of two peaks whose separation depends on the mean angle between the C—D bonds and the applied magnetic field, $\boldsymbol{B}$. It is a surprising fact that the time for the resonance process to take place is long relative to the times involved in random molecular motions and thus it is indeed the mean orientation which is measured. If the particular value of this splitting of the spectrum corresponds to the C—D bonds being at right angles to $\boldsymbol{B}$ then, for a deuterated methylene group in a hydrocarbon chain, this chain must lie in the direction of the field. If the magnetic field is arranged to be in the direction of the director, then splitting of the deuterium spectrum gives a direct measure of the mean departure of the axes of the molecules from the director. This statement is strictly only true if one considers orientation in the plane which divides the methylene groups in a symmetrical manner. Thus this spectral splitting can be used to measure the order parameter S associated with movement in this plane. Where one is dealing with hydrocarbon chains which rotate rapidly, the problem is slightly more complicated but, in the case of phospholipids, such rotation does not usually take place.

By deuterating particular parts of the molecules, these concepts can be extended to measure the mean orientation of particular parts of the hydrocarbon chains and it is thus possible to build up a picture of the occurrence of kinks in these chains and how these kinks develop with increasing temperature. For a further discussion of this technique, the reader is referred to the work by Silver [440] already referred to and original papers cited therein. A detailed discussion of this topic would be out of place in a general work of the kind offered here.

The general picture of lipid bilayers which emerges is one in which the head groups are relatively stationary and the hydrocarbon tails pack into an approximation to a two-dimensional hexagonal lattice. The degree of orientational order decreases as one proceeds towards the ends of the molecules remote from the head group and near the centre of the bilayer. As the temperature is increased, a phase change is eventually reached at which the degree of order within one monolayer suddenly decreases. It is interesting to note that, for a particular lipid, the temperature at which such a phase change takes place in a hydrated bilayer is nearly the same as the temperature at which an inflection corresponding to a phase change disappears in isotherms of monolayers studied at the air/water interface (Section 8.4). Though one should not build too much on this fact, it seems likely that the forces which stabilise the more ordered phase in a lipid bilayer act largely in one half of the bilayer only, and that interactions between the two halves are of small importance in this particular context. Studies of hydrated bilayers have been made using the differential scanning calorimeter and sharp peaks have been observed corresponding to the phase changes to which reference is made above, indicating that these phase changes are likely to be first order.

8.4 Lipids at the air/water interface

Over the last two decades, numerous studies of phospholipids at the air/water surface have been carried out. In general the isotherms observed have been similar to the one shown in Figure 8.4. In many cases, the curve to the left of the point of inflection appears to be nearly horizontal, which implies that that region corresponds to the coexistence

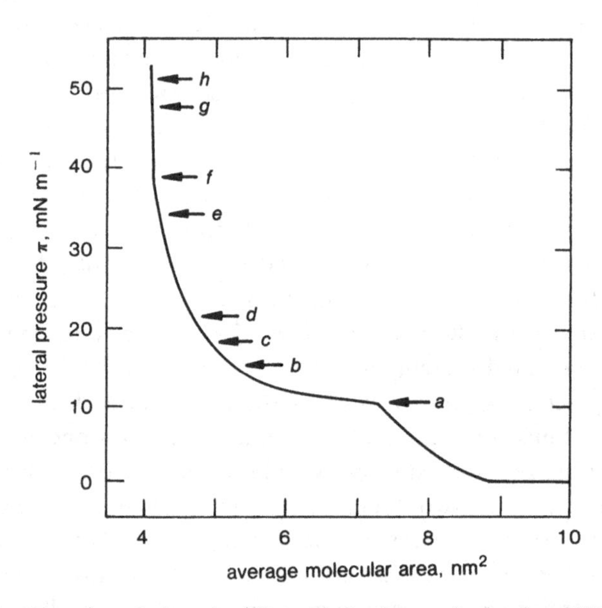

Figure 8.4. Dimyristoyl phosphatidic acid. Isotherm obtained at 23°C on a subphase containing 10^{-2} M NaCl and 5×10^{-5} M EDTA. The letters denote the points on the isotherm at which the diffraction peaks shown in Figure 8.5 were obtained. (This figure is reproduced from Helm, C.A., Möhwald, H., Kjaer, K. and Als-Nielsen, J. 1987 *Biophys. J.* **52** 381–90 by kind permission of the Biophysical Society and the authors.)

of two phases. As there exist a number of possible phospholipid head groups, each of which can be attached to many possible pairs of hydrocarbon chains, the range of materials which can be studied is very large and will not be explored in detail here.

However, since the mid-1980s new techniques have been introduced which give direct information about the structure of lipid layers at the air/water interface. The most important of these is the use of synchrotron X-ray radiation to obtain diffraction from the in-plane structure of such layers [79, 442–4]. Kjaer *et al.* [79] studied dimyristoyl phosphatidic acid (Figure 8.1) by this technique and also employed the fluorescence technique discussed in Section 3.3. A single pair of diffraction peaks on either side of the main beam associated with in-plane diffraction were obtained and their position varied in a systematic way as the surface pressure and hence area per molecule varied. The same group (Helm *et al.* [442]) subsequently carried out a more extensive study of this material and it is this study which will be discussed here. Figure 8.5 shows the diffraction peaks obtained at various points on the isotherm shown in Figure 8.4 and obtained at a temperature of 23 °C. These results, combined with the known cross sectional area of a hydrocarbon chain and results obtained by the fluorescence technique, lead to the following picture.

For a mean molecular area less than 0.77 nm^2, corresponding to a mean area per chain of half this value, a fluid phase coexists with a more highly ordered 'gel' phase. This latter phase corresponds to a hexagonal packing of the hydrocarbon chains. Such packing has been observed in electron diffraction studies of phospholipid bilayers made by Hui *et al.* [445] and phospholipid monolayers made by Fischer and Sackman [446] An ordered superlattice of the head groups would give rise to a half order reflection which Helm *et al.* [442] were unable to find and thus it is unlikely that the head groups are ordered. An analysis of the form of the diffraction peaks indicates that the so-called 'gel' phase is probably hexatic in nature. (The study made by Hui *et al.* [445] also supports the existence of a hexatic structure, though this paper predates the postulation of the hexatic phase.) From the experimental results, it is clear that the mean area per chain in the 'gel' phase is a function of pressure. Above a pressure, π_s, which corresponds to the diffraction pattern denoted by the letter f, there exists a truly condensed phase.

Helm *et al.* [443] subsequently studied dipalmitoyl phosphatidylcholine (DPPC). They concluded that the condensed two-dimensional phase normally involves a tilt of the hydrocarbon chains of about 30°

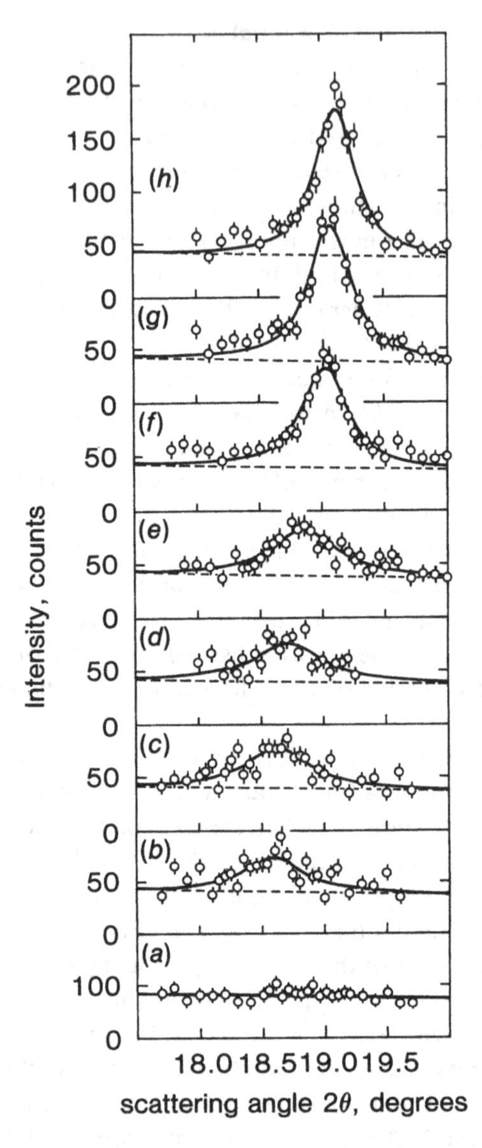

Figure 8.5. Diffraction peaks obtained from the film whose isotherm is shown in Figure 8.4. The synchrotron X-ray radiation used had a wavelength of 0.138 nm and was incident at a small downward angle on the film at the air/water interface and was diffracted through a horizontal angle, 2θ. (This figure is reproduced from Helm, C.A., Möhwald, H., Kjaer, K. and Als-Nielsen, J. 1987 *Biophys. J.* **52** 381–90 by kind permission of the Biophysical Society and the authors.)

but that this tilt decreases on further compression. Möhwald *et al.* [444] studied dimyristoyl phosphatidylethanolamine and showed that the liquid condensed phase was tilted and probably hexatic in nature.

Two recent studies of lipid monolayers have been made using neutron diffraction [447, 448]. In both cases the relative proportions of water and D_2O were varied and the analysis was based on the assumption that the interfacial region could be divided into several thin slabs, each having a particular value of refractive index to neutrons. Vaknin *et al.* [448] studied phosphatidylcholine. They showed that, in the liquid condensed phase, there was a chain tilt of about 33° and that the head group interpenetrates the water surface. About four water molecules are associated with each lipid molecule. Hunt *et al.* [449] and Mitchell *et al.* [450] studied DPPC and DMPA at the air/water interface by reflection FTIR spectroscopy and noted changes in the frequency of the CH_2 stretching bands at areas per molecule previously identified as regions at which phase changes take place. FTIR has not previously been applied in this particular context, but could become of value in providing a technique for the study of phase changes at the air/water interface which does not involve the sophisticated hardware needed for X-ray synchrotron studies.

In Section 3.3 the study of films at the air/water interface using fluorescent microscopy was discussed. Here, the fluorescent dye is squeezed out of more condensed phases but remains in those which are less condensed and thus regions of condensed material appear dark on a bright background. This technique has been particularly fruitful in the hands of Möhwald's group, who have used it in conjunction with the synchrotron X-ray diffraction technique and the study of isotherms to obtain a better understanding of lipid monolayers. In the two-phase region corresponding to the transition from the low density to the 'gel' region, there exists a structure consisting of many islands of 'gel', often arranged in a regular pattern as shown, for example, in Figure 8.6. Here DMPA containing 2.5 mM of an NBD dye is compressed from the point of inflection on the isotherm to about halfway through the two-phase region. In Figures 8.6(*e*) and (*f*) a regular two-dimensional crystalline structure may be observed. Figure 8.7 shows structures corresponding to a similar region of an isotherm of DMPC. Here it will be seen (particularly in (*e*) and (*f*)) that the 'gel' phase islands share a characteristic shape which arises from the nature of the intermolecular interaction. The edges of the domains nevertheless remain smooth. Miller and Möhwald [97] made use of the fact that lipids contain uncompensated dipoles and

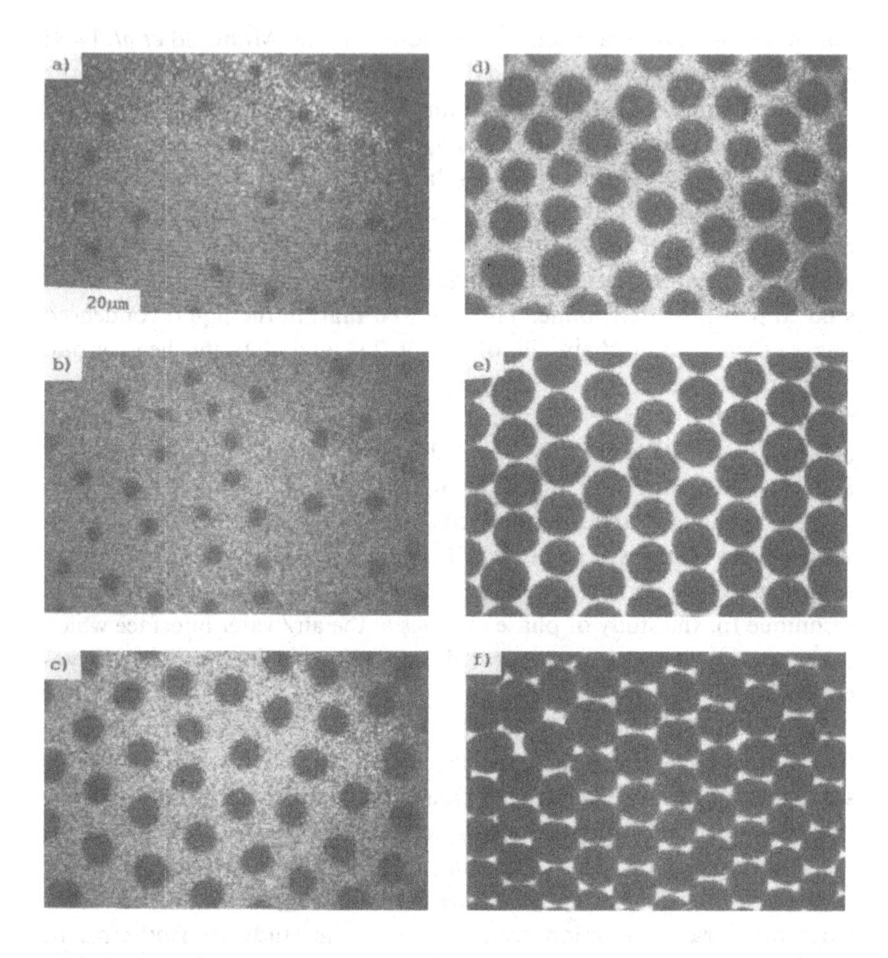

Figure 8.6. Fluorescence micrographs of a monolayer of L-α-dimyristoyl phosphatidic acid containing 2.5 mM of an NBD dye on increasing the surface pressure from the point of inflection on the isotherm, *a*, to about half way along the two-phase region. These results were obtained at a temperature of 10.5°C at pH 11.3 over a subphase containing 1 mM NaOH, 10 μM Na_2HPO_4, 100 mM NaCl, 2 μM $CaCl_2$ and 1 μM EDTA. (Reproduced from Lösche, M., Duwe, H.-P. and Möhwald, H. 1988 *J. Colloid Interface Sci.* **126** 432–44 by kind permission of the publishers and authors.)

are often charged to manipulate domains at the air/water interface using an electric field derived from a pointed electrode just above the water surface. This technique makes it possible to anchor a particular domain and study its evolution in detail. Figure 8.8 shows the growth of a domain of a more complex lipid (more details are given in the caption)

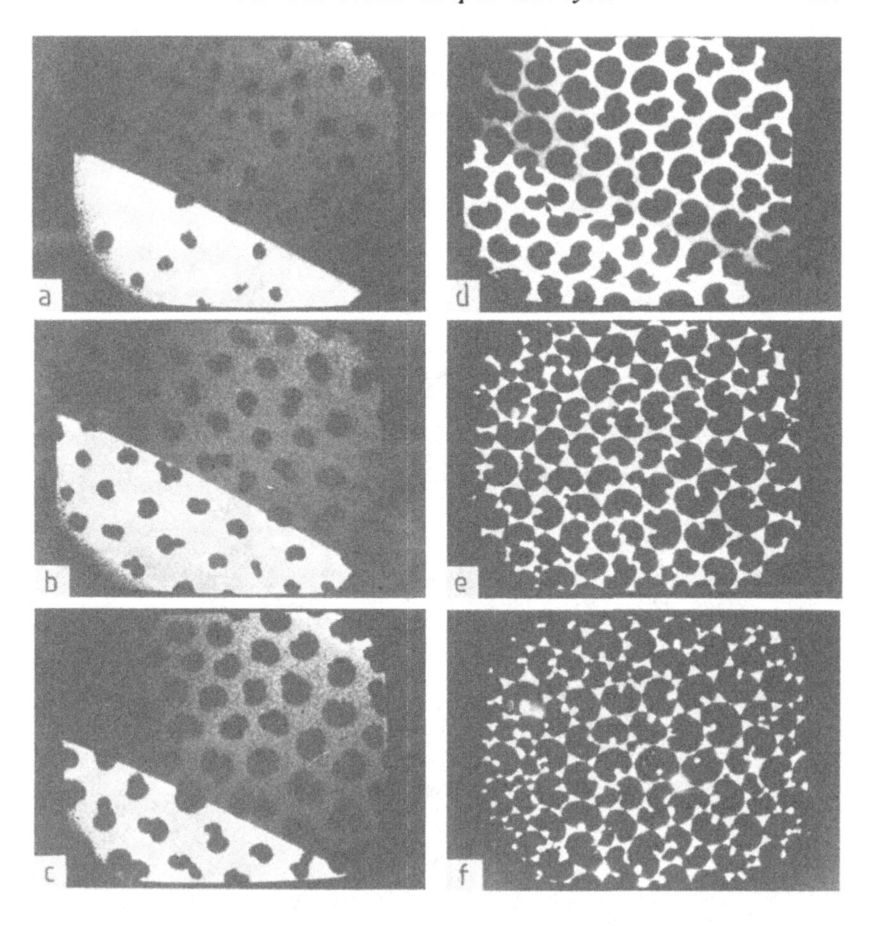

Figure 8.7. Fluorescence micrographs of a monolayer of L-α-dipalmitoyl phosphatidyl choline containing about 1% molar of an NBD dye at a temperature of 17°C and at pH 5.5. (Reproduced from Florsheimer, M. and Möhwald, H. 1989 *Chem. Phys. Lipids* **49** 231–41 by kind permission of the publishers and authors.)

which has *three* hydrocarbon tails. The outlines from (*c*) to (*f*) possess the self-similarity characteristic of fractals. Publications of this school of particular interest are [92, 97–9, 451–3].

8.5 The biotin–streptavidin system

The biotin–streptavidin affinity has been made use of by biochemists for some years but has only been used recently for the formation of ordered films. There exist two closely related proteins, avidin and streptavidin,

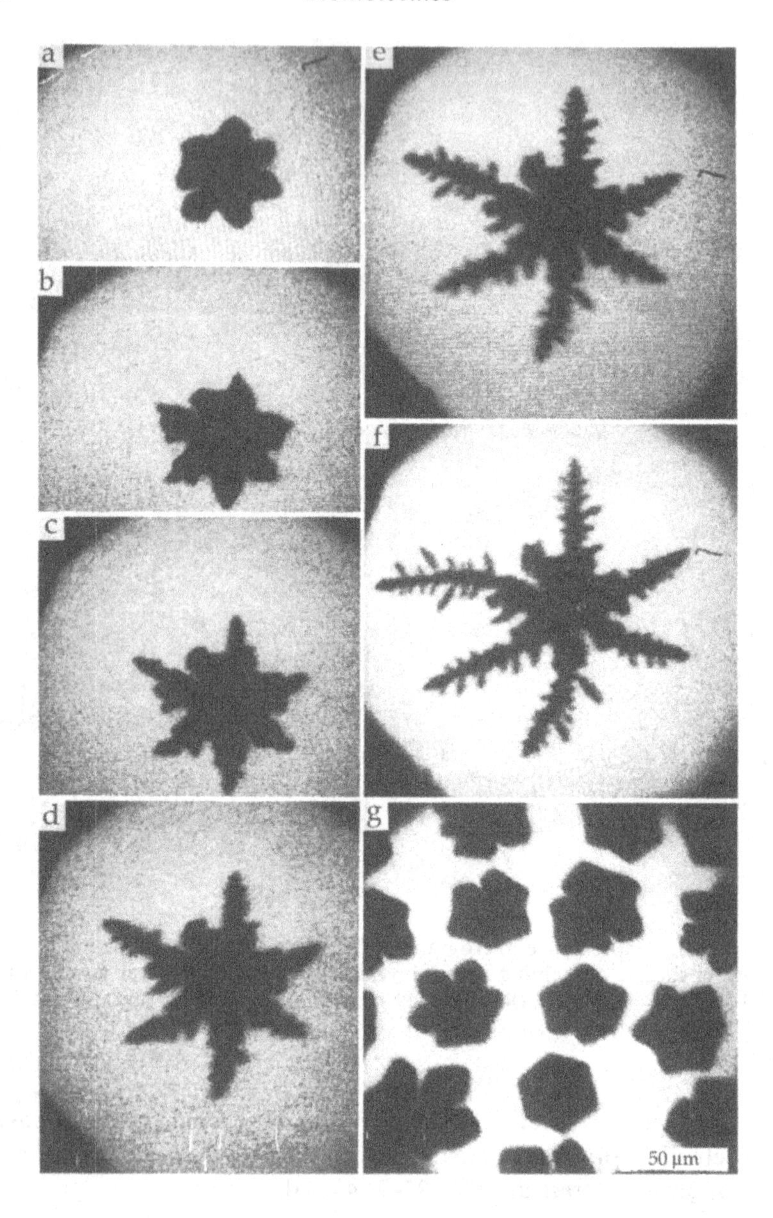

Figure 8.8. Fluorescence micrographs of a monolayer of 2-hexadecyl-2-(2-tetradecylpalmitoyl) glycero-3-phosphocholine containing about 1% molar of an amphiphilic dye probe taken at a temperature of 15°C. (Reproduced by kind permission of Professor Möhwald.)

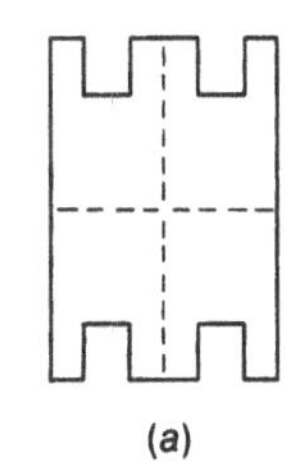

(a)

(b)

Figure 8.9. (*a*) Schematic diagram to illustrate the location of the binding sites on avidin and streptavidin molecules. (*b*) A lipid with a biotin head group. The biotin moiety is at the right of the diagram and locates in the cavities shown in a schematic manner in (*a*).

both of which consist of four sub-units, each of which is capable of binding one biotin molecule. The binding sites are arranged as shown in a schematic manner in Figure 8.9(*a*). In Figure 8.9(*b*) a lipid with a biotin head group is shown. Most of the work on this system has, in fact, made use of such lipids. The energy associated with this binding is about 1 eV, so the bond is relatively stable. Ringsdorf and his collaborators have made a series of studies of the reactions of streptavidin and biotinlipids at the air/water interface. The first of these (Ahlers *et al.* [454]) involved spreading the biotinlipid shown in Figure 8.9(*b*) at the air/water interface in a LB trough and injecting streptavidin into the subphase. The streptavidin was labelled with two fluorescent isothiocyanate molecules per protein. These authors were able to show that there was a specific interaction between the biotinlipid and the streptavidin at the air/water interface and that the two materials together formed a regular layer which they were able to study using the fluorescence microscopy techniques discussed in Section 8.4. The postulated structure of this layer is shown in Figure 8.10. They were also able to pick these layers up on an electron microscope grid (using an unspecified supporting film) and obtain electron micrographs which resolved individual streptavidin molecules. A

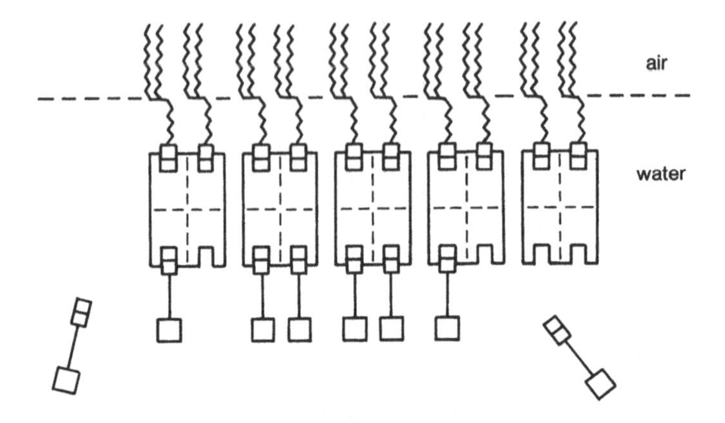

Figure 8.10. (*a*) Biotinlipid monolayer at the air/water interface. Streptavidin molecules are shown attached to this layer and a further layer of molecules, to which the biotin group has been attached, are shown in the process of binding to the streptavidin.

two-dimensional crystalline structure with a square cell was clearly visible. Thus, using this technique, it is possible to form good two-dimensional crystals of this protein with linear dimensions of the order 1 μ.

Ringsdorf's group has also investigated an interesting extension of this technique [455, 456]. It will be observed in Figure 8.10 that two unused binding sites per streptavidin molecule remain after an ordered layer of this material has reacted with the biotinlipid layer. It is thus possible to derivatise other molecules with the biotin group and hence, provided that these are water soluble, attach them to the underside of the streptavidin layer. Ahlers *et al.* [455] discuss how this process can be carried out using biotinylated ferritin, a large molecule containing iron, the presence of which can be easily detected. Herron *et al.* [456] studied a system consisting of a biotinlipid, streptavidin and a biotinylated antibody and produced two-dimensional crystals at the air/water interface.

Another important application of this interaction is in the formation of self-assembled bilayers on silver or gold surfaces. Ebersole *et al.* [457] showed that avidin and streptavidin molecules will adsorb onto clean layers of Au or Ag from an aqueous solution. Such layers can then be employed to capture biotinylated compounds. These authors used this technique to attach fragments of nucleic acid derived from the herpes virus to a solid support. A more popular approach to this general problem has, however, been to start by adsorption of a suitable biotin

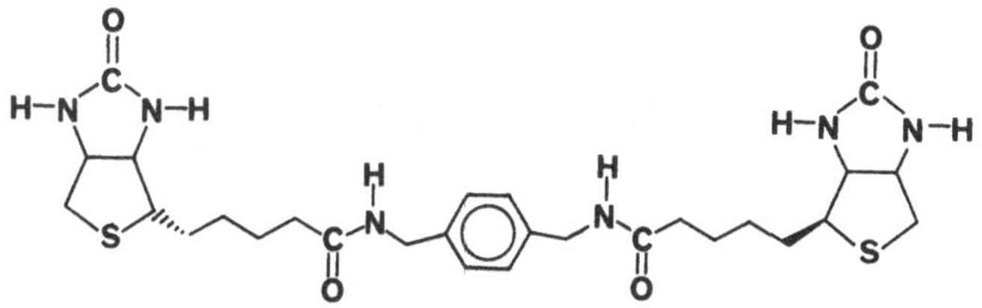

Figure 8.11. A rigid *bis*-biotinylated molecule as employed by Morgan *et al.* [462] to bring about formation of a self-assembled multilayer.

derivative onto a gold surface and then use this layer to capture avidin or streptavidin molecules. Examples of the application of this technique are given in Haeussling *et al.* [458, 459], Morgan *et al.* [460] and Taylor *et al.* [461]. In principle it should be possible use the upper, as yet unused, binding sites of the streptavidin to attach a molecule biotinylated at both ends and hence attach yet another streptavidin layer. Such a process could then be continued for a number of layers, as in the self-assembly process discussed in Section 6.5. Morgan *et al.* [462] report that their initial attempts to put such a process into practice were unsuccessful and attribute this failure to the fact that their *bis*-biotinylated compound was too long and flexible, so that it could couple to two streptavidin molecules in the same layer. However, when they used the shorter and stiffer *bis*-biotinylated compound, the structure of which is shown in Figure 8.11, they were successful and were able to build up a film containing nine protein monolayers. The ability to assemble multilayers of large molecules by this technique could prove to be an important step in the formation of true molecular electronic devices.

References

1. Worcester, G.R.G. 1966 *Sail and Sweep in China* (H.M.S.O., London).
2. Pockels, A. 1891 *Nature* **43** 437–9.
3. Tredgold, R.H. 1987 *Rep. Prog. Phys.* **50** 1609–56.
4. Franklin, B. 1774 *Phil. Trans. Roy. Soc.* **64** 445–60.
5. Pockels, A. 1892 *Nature* **46** 418–9.
6. Pockels, A. 1893 *Nature* **48** 152–4.
7. Pockels, A. 1894 *Nature* **50** 223–4.
8. Rayleigh, Lord. 1899 *Phil. Mag.* **48** 321–37.
9. Hardy, W.B. 1912 *Proc. Roy. Soc. A.* **86** 321–610.
10. Devaux, H. 1913 *Smithsonian Inst. Ann. Rep.* 261.
11. Langmuir, I. 1917 *J. Am. Chem. Soc.* **39** 1848–1906.
12. Langmuir, I. 1920 *Trans. Faraday Soc.* **15** 62-74.
13. Blodgett, K.B. 1935 *J. Am. Chem. Soc.* **57** 1007-22.
14. Gaines, G.L. 1966 *Insoluble Monolayers at Liquid-Gas Interfaces* (Interscience, New York).
15. Kuhn, H., Mobius, D. and Bucher, H. 1973 *Techniques of Chemistry*, Vol. 1, edited by A. Weissberger and B.W. Rositer (Wiley, New York) part III b.
16. Bigelow, W.C., Pickett, D. L. and Zisman, W.A. 1946 *J. Colloid Interface Sci.* **1** 513–18.
17. Netzer, L. and Sagiv, J. 1983 *J. Am. Chem. Soc.* **105** 674–6.
18. Agarwal, V.K. Igasaki, Y. and Mitsuhashi, H. 1976 *Thin Solid Films* **33** L31.
19. Reinitzer, F. 1888 *Mehft. Chem.* **9** 421–41.
20. Friedel, G. 1922 *Ann. Phys. (Paris)* **18** 273–474.
21. Isrealachvili, J.N. 1985 *Intermolecular and Surface Forces* (Academic Press, London).
22. London, F. 1930 *Z. Physik* **63** 245–79.
23. Casimer, H.B.G. and Polder, D. 1946 *Nature* **158** 787–8.
24. Young, T. 1805 *Phil. Trans. Roy. Soc.* **95** 74–87.
25. Zisman, W. 1964 in *Contact Angle, Wettability and Adhesion* edited by F. M. Fowkes, Advances in Chemistry Series, No. 43 (American Chemical Society, Washington).
26. De Gennes, P.G. 1985 *Rev. Mod. Phys.* **57** 827–63.
27. James, R.W. 1950 *The Optical Principles of the Diffraction of X-rays* (Bell, London).

28. Leadbetter, A.J. 1979 in *Molecular Physics of Liquid Crystals,* edited by G.R. Luckhurst and G.W. Gray (Academic Press, London) 285-316
29. Holley, C. and Bernstein, S. 1936 *Phys. Rev.* **49** 403.
30. Holley, C. and Bernstein, S. 1937 *Phys. Rev.* **52** 525.
31. Debye, P. 1913 *Verh. Deutsch. Phys. Ges.* **15** 678, 738, 857.
32. Debye, P. 1914 *Ann. Physik* **43** 49.
33. Waller, I. 1923 *Z. Physik* **17** 398.
34. Lesslauer, W. and Blasie, J. K. 1972 *Biophys. J.* **12** 175-90.
35. Tredgold, R.H., Allen, R.A. and Hodge, P. 1987 *Thin Solid Films* **155** 343-55.
36. Feigin, L.A. and Lvov, Y.M. 1988 *Makromol. Chem., Macromol. Symp.* **15** 259-74.
37. Belbeoch, B., Roulliay, M. and Tournarie, M. 1985 *C. R. Acad. Sci. Paris* Série II No. 12, 871-4.
38. Pomerantz, M. and Segmuller, A. 1980 *Thin Solid Films* **68** 33.
39. Srivastava, V.K. and Verma, A.R. 1966 *Solid State Commun.* **4** 367-71.
40. Matsuda, A., Sugi, M., Fukui, T., Iizima, S., Miyahara, M. and Otsubo, Y. 1977 *J. Appl. Phys.* **48** 771-4.
41. Mizushima, K., Nakayama, T. and Azuma, M. 1987 *Jpn. J. Appl. Phys.* Part I **26** 772-3.
42. Kamata, T., Umemura, J. and Takenaka, T. 1988 *Chem. Lett.* 1231-4.
43. Kajiyama, T., Hanada, I., Shuto, K. and Oishi, Y. 1989 *Chem. Lett.* 193-6.
44. Richardson, W. and Blasie, J.K. 1989 *Phys. Rev.* B.**39** 12 165-81.
45. Yang, K., Mu, J., Feng, X. and Chen, J. 1989 *Thin Solid Films* **178** 341-6.
46. Shiozawa, T. and Fukuda, K. 1989 *Thin Solid Films* **178** 421-5.
47. Chen, J. 1990 *Shandong Daxue Xuebao, Ziran Kexueban* **25** 98-103.
48. Ewald, P.P. 1913 *Physikal. Z.* **14** 465 and 1038.
49. von Laue, M. 1917 *Jahrbuch Radioakt Electronik* **11** 308.
50. Loretto, M.H. 1984 *Electron Beam Analysis of Materials.* (Chapman and Hall, London).
51. Willis, B.T.M. 1973 *Chemical Applications of Thermal Neutron Scattering* (Oxford University Press).
52. Fermi, E. 1950 *Nuclear Physics* (University of Chicago Press).
53. Hayter, J.B., Highfield, R.R., Pullman, B.J., Thomas, R.K., McMullen, A.I. and Penfold, J. 1981 *J. Chem. Soc. Faraday Trans.* I **77** 1437-48.
54. Nicklow, R.M., Pomerantz, M. and Segmuller, A. 1981 *Phys. Rev.* B **23** 1081-7.
55. Highfield, R.R., Thomas, R.K., Cummins, P.G., Gregory, D.P., Mingins, J., Hayter, J.B. and Scharpf, O. 1983 *Thin Solid Films* **99** 165-72.
56. Buhaenko, M.R., Grundy, M.J., Richardson, R.M. and Roser, S.J. 1988 *Thin Solid Films* **159** 253-65.
57. Grundy, M.J., Musgrove, R.J., Richardson, R.M. and Roser, S.J. 1990 *Langmuir* **6** 519-21.
58. Stroeve, P., Rabolt, J.F., Hilleke, R.O., Felcher, G.P. and Chen, S.H. 1990 *Mater. Res. Soc. Symp. Proc.* **166** 103-8.
59. Allara, D.L., Baca, A. and Pryde, C.A. 1978 *Macromolecules* **11** 1215-20.

60. Driscoll, W.G. and Vaughan, W. (editors) 1981 *Handbook of Optics,* (McGraw-Hill, New York).
61. Blight, L., Cumper, C.W.N. and Kyte, V.J. 1965 *J. Colloid Sci.* **20** 393-9.
62. Roberts, G.G., Vincett, P.S. and Barlow, W.A. 1981 *Phys. Technol.* **12** 69-75.
63. Kumehara, H., Kasuga, T., Watanabe, T. and Miyata, S. 1989 *Thin Solid Films* **178** 175-82.
64. Malcolm, B.R. and Davies, S.R. 1965 *J. Sci. Instrum.* **42** 359-60.
65. Langmuir, I. and Schaefer, V.J. 1937 *J. Am. Chem. Soc.* **59** 2400-14.
66. Buhaenko, M.R., Goodwin, J. W., Richardson, R.M. and Daniel, M.F. 1985 *Thin Solid Films* **134** 217-26.
67. Malcolm, B.R. 1985 *J. Colloid Interface Sci.* **104** 520-8.
68. Malcolm, B.R. 1985 *Thin Solid Films* **134** 201-8.
69. Daniel, M.F. and Hart, J.T.T. 1985 *J. Mol. Electron.* **1** 97-104.
70. Guyot, J. 1924 *Ann. Physique* **2** 505.
71. Frumkin, A.N. 1925 *Z. Physik. Chem. (Leipzig)* **166** 485.
72. Harkins, W.D. 1952 *The Physical Chemistry of Surface Films.* (Reinold, New York).
73. Pallas, N.R. and Pethica, B.A. 1985 *Langmuir* **1** 509-13.
74. Winch, P.J. and Earnshaw, J.C. 1989 *J. Phy. Condens. Matter* **1** 7187-205.
75. Moore, B.G., Knobler, C.M., Akamatsu, S. and Rondelez, F. 1990 *J. Phys. Chem.* **94** 4588-95.
76. Rettig, W. and Kuschel, F. 1990 *J. Colloid Interface Sci.* **140** 169-74.
77. Wostenholme, G.A. and Schulman, J.H. 1950 *Trans. Faraday Soc.* **46** 475-87.
78. Binks, B.P. 1991 *Adv. Colloid Interface Sci.* **34** 343-432.
79. Kjaer, K., Als-Nielsen, J., Helm, C.A., Laxhuber, L.A. and Möhwald, H. 1987 *Phys. Rev. Lett.* **58** 2224-7.
80. Richardson, R.M. and Roser, S.J. 1987 *Liq. Cryst.* **2** 797-814.
81. Dutta, P., Peng, J.B., Lin, B., Ketterson, J.B., Prakash, M., Georgopoulas, P. and Ehrlich, S. 1987 *Phys. Rev. Lett.* **58** 2228-31.
82. Bloch, J.M. Yun, W.B. Yang, X. Ramanathan, M., Montano, P.A. and Capasso, C. 1988 *Phys. Rev. Lett.* **61** 2941-4.
83. Kjaer, K., Als-Nielsen, J., Heim, C.A., Tippman-Krayer, P. and Möhwald, H. 1989 *J. Phys. Chem.* **93** 3200-6.
84. Lin, B., Peng, J.B., Ketterson, J.B. Dutta, P., Thomas, B.N., Buontempo, J. and Rice, S.A. 1989 *J. Chem. Phys.* **90** 2393-6.
85. Jacquemain, D., Wolf, S.G., Leveiller. F., Lahav, M., Leiserowitz, L., Deutsch, M., Kjaer, K. and Als-Nielsen, J. 1990 *J. Am. Chem. Soc.* **112** 7724-36.
86. Bohanon, T.M., Lin, B., Shih, M.C., Ice, G.E. and Dutta, P. 1990 *Phys. Rev.* B **41** 4846.
87. Lin, B. Bohanon, T.M. Shih, M.C. and Dutta, P. 1990 *Langmuir* **6** 1665-87.
88. Wolf, S.G., Leiserowitz, L., Lahav, M., Deutch, M. Kjaer, K. and Als-Nielsen, J. 1987 *Nature* **328** 63-6.
89. Kenn, R.M., Bohm, C., Bibo, A.M., Peterson, I.R., Möhwald, H., Als-Nielsen, J. and Kjaer, K. 1991 *J. Phys. Chem.* **95** 2092-7.
90. von Tscharner, V. and McConnell, H.M. 1981 *Biophys. J.* **36** 409-19.
91. Lösche, M. and Möhwald, H. 1984 *Rev. Sci. Instrum.* **55** 1968-72.

92. Lösche, M., Rabe, J., Fischer, A. Rucha, B.U. Knoll, W. and Möhwald, H. 1984 *Thin Solid Films* **117** 269–80.
93. Stine, K.J., Rauseo, S.A., Moore, B.G. Wise, J.A. and Knobler, C.M. 1990 *Phys. Rev.* A **41** 6884–92.
94. Berge, B., Simon, A.J. and Libchaber, A. 1990 *Phys. Rev.* A **41** 6893–900.
95. Fischer, A., Lösche, M., Möhwald, H. and Sackman, E. 1984 *J. Physique Lett.* **5** L785.
96. Helm, C.A., Laxhuber, L.A. Lösche, M. and Möhwald, H. 1986 *Colloid Polym. Sci.* **264** 1.
97. Miller, A. and Möhwald, H. 1986 *Europhys. Lett.* **2** 67–74.
98. Miller. A., Helm, C.A. and Möhwald, H. 1987 *J. Physique.* **48** 693–701.
99. Vanderlick, T.K. and Möhwald, H. 1990. *J. Phys. Chem.* **94** 886–90.
100. Mandelbrot, B.B. 1982 *The Fractal Geometry of Nature* (Freeman, Oxford).
101. Keller, D.J., Korb, J.P. and McConnell, H.M. 1987 *J. Phys. Chem.* **91** 6417–22.
102. McConnell, H.M. and Moy. V.T. 1988 *J. Phys. Chem.* **92** 4520–5.
103. McConnell, H.M. 1990 *J. Phys. Chem.* **94** 4728–31.
104. Rasing, T. and Shen, Y.R., Kim, M.W. Grubb, S. and Bock, J. 1985 *Phys. Rev. Lett.* **55** 2903–6.
105. Rasing, T. and Shen, Y.R., Kim, M.W., Grubb, S. and Bock, J. 1985 *Springer Ser. Opt. Sci.; Laser Spectrosc.* **49** 307–10.
106. Boyd, G.T., Shen, Y.R. and Hansch, T.W. 1985 *Springer Ser. Opt. Sci.; Laser Spectrosc.* **49** 322–3.
107. Boyd, G.T., Shen, Y.R. and Hansch, T.W. 1986 *Optics Lett.* **11** 97–9.
108. Kajikawa, K., Shirota, K., Takezoe, H. and Fakuda, A. 1990 *Jpn. J. Appl. Phys.* Part 1 **29** 913–17.
109. Shirota, K., Kajikawa, K., Takezoe, H. and Fakuda, A. 1990 *Jpn. J. Appl. Phys.* Part 1 **29** 750–5.
110. Vogel, V., Mullin, C.S., Shen, Y.R. and Kim, M.W. 1990 *Mater. Res. Soc. Symp. Proc.* **177** 363–5.
111. Zhao, X., Subrahmanyan, S. and Eisenthal, K.B. 1990 *Chem. Phys. Lett* **177** 558–62.
112. Stallberg-Stenhagen, S. and Stenhagen, E. 1945 *Nature* **155** 239–40.
113. Bibo, A.M., Knobler, C.M. and Peterson, I.R. 1991 *J. Phys. Chem.* **95** 5591–9.
114. Nelson, D.R. and Halperin, B.I. 1979 *Phys. Rev.* B **19** 2457–84.
115. Brock, J.D., Birgeneau, R.J., Litser, J.D. and Aharony, A. 1989 *Phys. Today* July 52–9.
116. Peterson, I.R. 1986 *J. Mol. Electron.* **2** 95–9.
117. Bibo, A.M. and Peterson, I.R. 1989 *Thin Solid Films* **178** 81–92.
118. Tredgold, R.H. and Jones, R. 1989 *Langmuir* **5** 531–3.
119. Roberts, G.G. 1990 *Langmuir Blodgett Films* (Plenum Press, New York).
120. Handy, R.M. and Scala, L.C. 1966 *J. Electrochem. Soc.* **113** 109–16.
121. Mann, B. and Kuhn, H. 1971 *J. Appl. Phys.* **42** 4398–405,
122. Gundlach, K.H. and Kadlech, J. 1974 *Chem. Phys. Lett.* **25** 293–5.
123. Polymeropoulos, E.E. 1977 *J. Appl. Phys.* **48** 2404–7.
124. Polymeropoulos, E.E. 1978 *Solid State Commun.* **28** 883–5.
125. Polymeropoulos, E.E. and Sagiv, J. 1978 *J. Chem. Phys.* **79** 1836–47.

126. Tredgold, R.H., Jones, R., Evans, S.D. and Williams, P.I. 1986 *J. Mol. Electron.* **2** 147–9.
127. Peterson, I.R. 1980 *Aust. J. Chem.* **33** 1713–6.
128. Procarione, W.L. and Kauffman, J.W. 1974 *Chem. Phys. Lipids* **12** 251–60.
129. Tredgold, R.H. and Winter, C.S. 1981 *J. Phys. D: Appl. Phys.* **14** L185–8.
130. Peterson, I.R. 1986 *J. Mol. Electron.* **2** 95–9.
131. Peterson, I.R. 1984 *Thin Solid Films* **116** 357–66.
132. Peterson, I.R. 1987 *J. Mol. Electron.* **3** 103.
133. Bibo, A.M. and Peterson, I.R. 1989 *Thin Solid Films* **178** 81–92.
134. Kitaigorodskii, A.I. 1961 *Organic Chemical Crystallography* (Consultants Bureau, New York).
135. Robinson, I., Sambles, J.R. and Peterson, I.R. 1989 *Thin Solid Films* **172** 149–58.
136. Russell, G.J., Petty, M.C., Peterson, I.R., Roberts, G.G. Lloyd, J.P. and Kan, K.K. 1984 *J. Mater. Sci.* **3** 25–8.
137. Garoff, S. Deckman, H.V., Dunsmuir, J.H. Alvarez, M.S. and Bloch, J.M. 1986 *J. Physique* **47** 701–9.
138. Peterson, I.R., Steitz, R. Krug, H. and Voigt-Martin, I. 1990 *J. Physique* **51** 1003–26.
139. Bonnerot, A., Chollet, P.A. Frisby, H. and Hoclet, M. 1985 *Chem. Phys.* **97** 365–77.
140. Robinson, I., Peterson, I.R. and Sambles, J.R. 1990 *Phil. Mag. Lett.* **62** 101–106.
141. Donnay, J.D.H. and Ondik, H.M. 1972 *Crystal Data Determinative Tables*, third edition, Vol. I (U.S. Department of Commerce National Bureau of Standards, Washington D.C.).
142. Matsuda, H., Kishi, E., Kuroda, R., Albrecht, O., Eguchi, K. Hatanaka, K. and Nagagari, T. 1993 *Thin Solid Films*. In the press.
143. Nakahama, H., Miyata, S., Wang, T.T. and Tasaka, S. 1986 *Thin Solid Films.* **141** 165–9.
144. Naselli, C., Swalen, J.D. and Rabolt, J.F. 1989 *J. Chem. Phys.* **90** 3855–60.
145. Jones, R., Tredgold, R.H., Ali-Adib, Z., Dawes, A.P.L. and Hodge, P. 1991 *Thin Solid Films* **200** 375–84.
146. Prakash, M., Peng, J.B., Ketterson, J.B. and Dutta, P. 1987 *Thin Solid Films* **146** L15–17.
147. Ohe, H. Tajima, K. and Sano, H. 1987 *Nippon Kagaka Kaishi* (11) 2070–5.
148. Heeseman, J. 1980 *J. Am. Chem. Soc.* **102** 2167–81.
149. Nakahara, H. and Fukuda, K. 1983 *J. Colloid and Interface Sci.* **93** 530–9.
150. Blinov, L.M., Dubinin, N.V., Mikhnev, L.V. and Yudin, S.G. 1984 *Thin Solid Films.* **120** 161–70.
151. Jones, R., Tredgold, R.H., Hoorfar, A., Allen, R.A. and Hodge, P. 1985 *Thin Solid Films* **134** 57–66.
152. Kawai, T., Umemura, J. and Takenaka, T. 1989 *Langmuir* **5** 1378–83.
153. Kawai, T., Umemura, J. and Takenaka, T. 1990 *Langmuir* **6** 672–6.
154. Tanaka, M., Ishizuka, Y., Matsumoto, M., Nakamura, T., Yabe, A. Nakanishi, H., Kawabata, Y., Takahashi, H., Tamura, S., Tagaki, W., Nakahara, H. and Fukuda, K. 1987 *Chem. Lett.* 1307–10.

155. Yabe, A., Kawabata, Y., Niino, H., Tanaka, M., Ouchi, A., Takahashi, H., Tamura. S., Tagaki, W., Nakahara, H. and Fukuda, K. 1988 *Chem. Lett.* 1-4.
156. Nishiyama, K. and Fujihira, M. 1988 *Chem. Lett.* 1257-60.
157. Nakahara, H., Fukuda, K., Shimomura, M. and Kunitake, T. 1988 *Nippon Kagaka Kaishi* 1001-10.
158. Tachibana, H., Goto, A., Nakamura, T., Matsumoto, M., Manda, E., Niino, H., Yabe, A. and Kawabata, Y. 1989 *Thin Solid Films* **179** 207-13.
159. Tachibana, H. Nakamura, T., Matsumoto, M. Komizu, H., Manda, E., Niino, H., Yaba, A. and Kawabata, Y. 1989 *J. Am. Chem. Soc.* **111** 3080-1.
160. Liu, Z., Loo, B.H., Baba, R. and Fujishima, A. 1990 *Chem. Lett.* 1023-26.
161. Barnik, M.I., Kozenkov, V.M., Shtykov, N.M., Palto, S.P. and Yudin, S.G. 1989 *J. Mol. Electron* **5** 53-6.
162. Liu, Z.F., Loo, B.H., Hashimoto, K., and Fujishima, A. 1991 *J. Electroanal. Chem. Interfacial Electrochem.* **297** 133-44.
163. Falk, J.E. 1975 edited by K.M. Smith *Porphyrins and Metalloporphyrins,* (Elsevier, Amsterdam).
164. Alexander, A.E. 1937 *J. Chem. Soc.* 1813.
165. Bergeron, J.A., Gaines, G.L. and Bellamy, W.D. 1967 *J. Colloid Interface Sci.* **25** 97-106.
166. Jones, R., Tredgold, R.H. and Hodge, P. 1983 *Thin Solid Films* **99** 25-32.
167. Jones, R., Tredgold, R.H., Hoorfar, A. and Hodge, P. 1984 *Thin Solid Films* **113** 115-28.
168. Jones, R., Tredgold, R.H. and Hoorfar, A. 1985 *Thin Solid Films* **123** 307-14.
169. Luk. S.Y., Mayers, F.R. and Williams, J.O. 1988 *Thin Solid Films* **157** 69-79.
170. van Galen, D.A. and Majda, M. 1988 *Anal. Chem.* **60** 1549-53.
171. Nagamura, T., Kamata, S., Sakai, K. Matano, K. and Ogawa, T. 1989 *Thin Solid Films* **179** 293-6.
172. Baker, S. Roberts, G.G. and Petty, M.C. 1983 *IEE Proc.* I **130** 260-3.
173. Cook, M.J., Daniel, M.F., Harrison, K.J., McKeown, N.B. and Thomson, A.J. 1987 *J. Chem. Soc., Chem. Commun.* 1086-8.
174. Cook, M.J., Daniel, M.F., Harrison, K.J., McKeown, N.B. and Thomson, A.J. 1987 *J. Chem. Soc., Chem. Commun.* 1148-50.
175. Cook, M.J., McKeown, N.B., Thomson, A.J., Harrison, K.J., Richardson, R.M., Davis, A.N. and Roser, S.J. 1989 *Chem. Mater.* **1** 287-9.
176. McKeown, N.B. Cook, M.J. Thomson, A.J. Harrison, K.J. Daniel, M.F., Richardson, R.M. and Roser, S.J. 1988 *Thin Solid Films* **159** 469-78.
177. Bull, R.A. and Bulkowski, J.E. 1983 *J. Colloid Interface Sci* **92** 1-12.
178. Baker, S., Petty, M.C., Roberts, G. G. and Twigg, M. V. 1983 *Thin Solid Films* **99** 53-9.
179. Hua, Y.L., Roberts, G.G., Ahmad, M.M. Petty, M.C., Hanack, M. and Rein, M. 1986 *Phil. Mag.* B **53** 105-113.
180. Fujiki, M. Tabei, H. and Kurihara, T. 1988 *J. Phys. Chem.* **92** 1281-5.

181. Nakahara, H., Fukuda, K., Katahara, K. and Nishi. H. 1989 *Thin Solid Films* **178** 361-6.
182. Ogawa, K., Yonehara, H., Shoji, T., Kinoshita, S., Maekawa, E. Nakahara, H. and Fukuda, K. 1989 *Thin Solid Films* **178** 439-43.
183. Nichogi, K., Waragai, K., Taomoto, A., Saito, Y. and Asakawa, S. 1989 *Thin Solid Films* **179** 297-301.
184. Fryer, J.R., McConnell, C.M., Hann, R.A., Eyres, B.L. and Gupta, S.K. 1990 *Phil. Mag.* B **61** 843-52.
185. Brynda, E., Kalvoda, L., Koropecky, I., Nespurek, S. and Rakusan, J. 1990 *Synth. Met.* **37** 327-33.
186. Brynda, E., Koropecky, I., Kalvoda, L. and Nespurek, S. 1991 *Thin Solid Films* **199** 375-84.
187. Fukui, M., Katayama, N., Ozaki, Y., Araki, T. and Iriyama, K. 1991 *Chem. Phys. Lett.* **177** 247-51.
188. Pace, M.D., Barger, W.R. and Snow, A.W. 1989 *Langmuir* **5** 973-8.
189. Shutt, J.D., Batzel, D.A., Sudiwala, R.V., Rickert S.E. and Kenney, M.E. 1988 *Langmuir* **4** 1240-7.
190. Shutt, J.D. and Rickert, S.E. 1989 *J. Mol. Electron.* **5** 129-34.
191. Liu, Y., Shigehara, K. and Yamada, A. 1989 *Thin Solid Films* **179** 303-8.
192. Ruaudel-Teixier, A., Barraud, A., Belbeoch, B. and Roulliay, M. 1983 *Thin Solid Films* **99** 33-40.
193. Lesieur, P., Vandevyver, M, Ruaudel-Texier, A. and Barraud, A. 1988 *Thin Solid Films* **159** 315.
194. Palacin, S., Lesieur, P., Stefanelli, I. and Barraud, A. 1988 *Thin Solid Films* **159** 83.
195. Palacin, S. and Barraud, A. 1989 *J. Chem. Soc., Chem. Commun* 45-7.
197. Ulman, A. 1991 *Ultrathin Organic Films* (Academic Press, Boston).
198. Cemel, A., Fort, T. and Lando, J.B. 1972 *J. Polymer Sci.* Part A-1 **10** 2061-83.
199. Naegele, D., Lando, J.B. and Ringsdorf, H. 1977 *Macromolecules* **10** 1339-44.
200. Rabe, J.P., Rabolt, J.F. Brown, C.A. and Swalen, J.D. 1982 *J. Chem. Phys.* **84** 4096-102.
201. Laschewsky, A., Ringsdorf, H. and Schmidt, G. 1985 *Thin Solid Films* **134** 153-72.
202. Laschewsky, A., Ringsdorf, H. and Schmidt, G. 1988 *Polymer* **29** 448-56.
203. Laschewsky, A. and Ringsdorf, H. 1988 *Macromolecules* **21** 1936-41.
204. Barraud, A. Rosilio, C. and Ruaudel-Teixier, A. 1977 *J. Colloid Interface Sci.* **62** 509-23.
205. Barraud, A., Rosilio, C. and Ruaudel-Teixier, A. 1980 *Thin Solid Films* **68** 91-8.
206. Uchida, M., Tanizaki, T., Kunitake, T. and Kajiyama, T. 1989 *Macromolecules* **22** 2381-7.
207. Uchida, M., Tanizaki, T., Oda, T. and Kajiyama, T. 1991 *Macromolecules* **24** 3238-43.
208. Bloor, D. and Chance, R.R. 1985 *Polydiacetylenes, Synthesis, Structure and Electronic Properties.* (Martinus Nijhoff, Dordrecht/Boston/ Lancaster).
209. Tieke, B. Graf, H-J., Wegner, G., Naegele, D., Ringsdorf, H.,

Banerjie, A., Day, D. and Lando, J.B. 1977 *Colloid Polymer Sci.* **255** 521–31.

210. Tieke, B., Wegner, G., Naegele, D. and Ringsdorf, H. 1976 *Angew. Chem. Int. Ed. Engl.* **15** 764.
211. Tieke, H., Lieser, G. and Wegner, G. 1979 *J. of Polymer Sci.: Polymer Chem. Ed.* **17** 1631-44.
212. Tieke, B. Enkelmann, V. Kapp, H., Lieser, G. and Wegner, G. 1981 *J. Macromol. Sci. - Chem.* A **15** 1045–58.
213. Tieke, B. and Weiss, K. 1984 *J. Colloid Interface Sci.* **101** 129–48.
214. Tomioka, Y., Tanaka, N. and Imazeki, S. 1989 *J. Chem. Phys.* **91** 5694–700.
215. Tamura. H., Mino, N. and Ogawa, K. 1989 *Thin Solid Films* **179** 33–9.
216. Fukuda, A., Koyama, T., Hanabusa, K., Shirai, H., Nakahara., H. and Fukuda, K. 1988 *J. Chem. Sci., Chem. Commun.* 1104–6.
217. Ahmed, F. R., Wilson, E.G. and Moss, G.P. 1990 *Thin Solid Films* **187** 141–53.
218. Tredgold, R. H. and Winter, C.S. 1982 *J. Phys. D: Appl. Phys.* **15** L55–8.
219. Tredgold, R.H. and Winter, C.S. 1983 *Thin Solid Films* **99** 81–5.
220. Winter, C.S. and Tredgold, R.H. 1983 *I.E.E. Proc.* Pt 1 **130** 256–9.
221. Winter, C.S., Tredgold, R.H., Hodge, P. and Khoshdel, E. 1984 *IEE. Proc.* Pt 1 **131** 125–8.
222. Tredgold, R.H., Vickers, A.J., Hoorfar, A., Hodge, P. and Khoshdel, E. 1985 *J. Phys. D: Appl. Phys.* **18** 1139–45.
223. Tredgold, R.H. and El-Badawy, Z.I. 1985 *J. Phys. D: Appl. Phys.* **18** 2483–7.
224. Hodge, P., Khoshdel, E., Tredgold, R.H., Vickers, A. and Winter, C.S. 1985 *British Polymer J.* **17** 368–71.
225. Tredgold, R.H., Allen, R.A., Hodge, P. and Khoshdel, E. 1987 *J. Phys. D: Appl. Phys.* 1385–8.
226. Tredgold, R.H. 1987 *Thin Solid Films* **152** 223–30.
227. Tredgold, R.H., Young, M.C.J., Hodge, P. and Khoshdel, E. 1987 *Thin Solid Films* **151** 441–9.
228. Jones, R., Winter, C.S., Tredgold, R.H. Hodge, P. and Hoorfar, A. 1987 *Polymer* **28** 1619–26.
229. Young, M.C.J., Jones, R., Tredgold, R.H. Lu, W.X., Ali-Adib, Z. Hodge, P. and Abbasi, F. 1989 *Thin Solid Films* **182** 319–32.
230. Hodge, P., Davis, F. and Tredgold, R.H. 1990 *Phil. Trans. Roy. Soc. Lond.* A **330** 153–66.
231. Jones, R., Tredgold, R.H. Davis, F. and Hodge, P. 1990 *Thin Solid Films* **186** L51–4.
232. Ali-Adib, Z., Tredgold, R.H., Hodge, P. and Davis, F. 1991 *Langmuir* **7** 363–6.
233. Elbert, R., Laschewsky, A. and Ringsdorf, H. 1985 *J. Am. Chem. Soc.* **107** 4134–41.
234. Biddle, M.B., Lando, J.B., Ringsdorf, H., Schmidt, G. and Schneider, J. 1988 *Colloid Polymer Sci.* **266** 806–13.
235. Erdelen, C., Laschewsky, A., Ringsdorf, H., Schneider, J. and Schuster, A. 1989 *Thin Solid Films* **180** 153–66.
236. Schneider, J., Erdelen, C., Ringsdorf, H. and Rabolt, J.F. 1989 *Macromolecules* **22** 3475–80.

237. Penner, T.L., Schildkraut, J.S., Ringsdorf, H. and Schuster, A. 1991 *Macromolecules* **24** 1041–9.
238. Gabrielli, G., Puggelli, M. and Ferroni, E. 1974 *J. Colloid Interface Sci.* **47** 145.
239. Mumby, S.J., Swalen, J.D. and Rabolt, J.F. 1986 *Macromolecules* **19** 1054–9.
240. Naito, K. 1989 *J. Colloid Interface Sci.* **131** 218–25.
241. Brinkhuis, R.H.G. and Schouten, A.J. 1991 *Macromolecules* **24** 1487–95 and 1496–504.
242. Kawaguchi, T., Nakahara, H. and Fukuda, K. 1985 *J. Colloid Interface Sci.* **104** 290–3.
243. Schoondorp, M.A. Vorenkamp, E. and Schouten, A.J. 1991 *Thin Solid Films* **196** 121–36.
244. Oguchi, K. Yoden, T., Sanui, K. and Ogata, N. 1986 *Polymer J.* **18** 887–90.
245. Oguchi, K., Yoden, T., Kosaka, Y., Watanabe, M., Sanui, K. and Ogata, N. 1988 *Thin Solid Films* **161** 305–13.
246. Watanabe, M., Kosaka, Y., Oguchi, K., Sanui, K. and Ogata, N 1988 *Macromolecules* **21** 2997–3003.
247. Lupo, D., Prass, W. and Scheunemann, U. 1989 *Thin Solid Films* **178** 403–11.
248. Nerger, D., Ohst, H., Schopper, H-C. and Wehrmann, R. 1989 *Thin Solid Films* **178** 253–9.
249. He, P., Bai, J., Yao, G., Zhou, G. and Wang, C. 1989 *J. Mater. Sci.* **24** 1901–3.
250. Iyoda, T., Ando, M., Kaneko, T., Ohtani, A., Shimidzu, T. and Honda, K. 1986 *Tetrahedron Lett* **27** 5633–6.
251. Yang, X.Q., Chen, J., Hale, P.D., Inagaki, T., Skotheim, T.A. Fischer, D.A., Okamoto, Y., Samuelson, L., Tripathy, S., Hong, K., Watanabe, I., Rubner, M.F. and den Boer, M.L. 1989 *Langmuir* **5** 1288–92.
252. Carr, Neil., Goodwin, M.J., McRoberts, A.M., Gray, G.W., Marsden, R. and Scrowston, R.M. 1987 *Makromol. Chem. Rapid Commun.* **8** 487–93.
253. Tamura, M., Ishida, H. and Sekiya, A. 1988 *Chem. Lett.* 1277–80.
254. Sekiya, A. and Tamura, M. 1990 *Chem. Lett.* 707–10.
255. Kakimoto, M., Suzuki, M., Konishi, T. Imai, Y., Iwamoto, M. and Hino, T. 1986 *Chem. Lett.* 823–6.
256. Uekita, M., Awaji, H. and Murata, M. 1988 *Thin Solid Films* **160** 21–32.
257. Nishikata, Y., Kakimoto, M. and Imai, Y. 1988 *J. Chem. Soc., Chem. Commun.* 1040–2.
258. Era, M., Shinozaki, M., Tokito, S., Tsutsui, T. and Saito, S. 1983 *Chem. Lett.* 1097–1100.
259. Nishikata, Y., Kakimoto, M., Morikawa, A., Kobayashi, I., Imai, Y., Hirata, Y., Nishiyama, K. and Fujihara, M. 1989 *Chem. Lett.* 861–4.
260. Baker, S., Seki, A. and Seto, J. 1989 *Thin Solid Films* **180** 263–70.
261. Sorita, T., Miyake, S., Fujioka, H. and Nakajima, H. 1991 *Jpn. J. Appl. Phys.* Part 1 **30** 131–5.
262. Fujiwara, I., Ishimoto, C. and Seto, J. 1991 *J. Vac. Sci. Technol.* B **9** (2 Part 2) 1148–53.

263. Ito, S., Kanno, K., Ohmori, S., Onogi, Y. and Yamamoto, M. 1991 *Macromolecules* **24** 659-65.
264. Nishikata, Y., Kakimoto, M. and Imai, Y. 1989 *Thin Solid Films* **179** 191-7.
265. Era, M., Kamiyama, K., Yoshiura, K., Momii, T., Murata, H., Tokito, S., Tsutsui, T. and Saito, S. 1989 *Thin Solid Films* **179** 1-8.
266. Kamiyama, K., Era, M., Tsutsui, T. and Saito, S. 1990 *Jpn. J. Appl. Phys.* Part 2 **29** L840-2
267. Pauling, L., Corey, R.B. and Branson, H.R. 1951 *Proc. Nat. Acad. Sci. U.S.A.* **37** 205.
268. Malcolm, B.R. 1962 *Nature* **195** 901-2
269. Malcolm, B.R. 1966 *Polymer* **7** 595-602.
270. Malcolm, B.R. 1968 *Biochem. J.* **110** 733-7.
271. Malcolm, B.R. 1971 *J. Polymer Sci.* Part C **34** 87-99.
272. Malcolm, B.R. 1985 *J. Colloid Interface Sci.* **104** 520-9.
273. Takenaka, T., Harada, K. and Matsumoto, M. 1980 *J. Colloid Interface Sci.* **73** 569-77.
274. Takeda, F., Matsumoto, M., Takenaka, T. and Fujiyoshi, Y. 1981 *J. Colloid Interface Sci.* **84** 220-7.
275. Takeda, F., Matsumoto, M., Takenaka, T., Fujiyoshi, Y. and Uyeda, N. 1983 *J. Colloid Interface Sci.* **91** 267-71.
276. Winter, C.S. and Tredgold, R.H. 1985 *Thin Solid Films* **123** L1-3.
277. Jones, R. and Tredgold, R.H. 1988 *J. Phys. D: Appl. Phys.* **21** 449-53.
278. Hickel, W., Duda, G., Jurich, M., Kroehl, T., Rochford, K., Stegeman, G.I., Swalen, J.D., Wegner, G. and Knoll, W. 1990 *Langmuir* **6** 1403-7.
279. Joyner, R.D. and Kenney, M.E. 1960 *J. Am. Chem. Soc.* **82** 5790
280. Fryer, J.F. and Kenney, M.E. 1988 *Macromolecules* **21** 259-62.
281. Orthmann, E. and Wegner, G. 1986 *Makromol. Chem., Rapid Commun.* **7** 243-7.
282. Orthmann, E. and Wegner, G. 1986 *Angew. Chem. Int. Ed. Engl.* **25** 1105-7
283. Caseri, W., Sauer T. and Wegner, G. 1988 *Makromol. Chem. Rapid Commun.* **9** 651-7.
284. Rabe, J. P., Sano, M., Batchelder, D. and Kalachev, A.A. 1988 *J. Microscopy* **152** 573-83.
285. Sauer, T., Arndt, T., Batchelder, D., Kalachev, A.A. and Wegner, G. 1990 *Thin Solid Films* **187** 357-74.
286. Kalachev, A.A., Sauer, T., Vogel, V., Plate, N.A. and Wegner,G. 1990 *Thin Solid Films* **188** 341-53.
287. Crockett, R.G.M., Campbell, A.J. and Ahmed, F.R. 1990 *Polymer* **31** 602-8.
288. Girling, I.R., Kolinsky, P.V., Cade, N.A., Earls, J.D. and Peterson, I.R. 1985 *Optics Commun* **55** 289-92.
289. Girling, I.R., Cade, N.A., Kolinsky, P.V., Earls, J.D. Gross, G.H. and Peterson, I.R. 1985 *Thin Solid Films* **132** 101-112.
290. Neal, D. B., Petty, M.C., Roberts, G.G., Ahmad, M.M., Feast, W.J., Girling, I.R., Cade, N.A., Kolinsky, P.V. and Peterson, I.R. 1986 *Electron. Lett.* **22** 460-2.
291. Neal, D.B., Petty, M.C., Roberts, G.G., Ahmad, M.M., Feast, W.J.,

Girling, I.R., Cade, N.A., Kolinsky, P.V. and Peterson, I.R. 1986 *Proceedings of the Sixth IEEE International Symposium on Applications of Ferroelectrics June 8-11 1986.*

292. Girling, I.R., Cade, N.A., Kolinsky, P.V., Jones, R.J., Peterson, I.R., Ahmad, M.M., Neal, D.B., Petty, M.C., Roberts, G.G. and Feast, W.J. 1987 *J. Opt. Soc. Am. B; Opt. Phys.* **4** 950-5.
293. Tredgold, R.H., Young, M.C.J. Jones, R., Hodge, P., Kolinsky, P.V. and Jones, R.J. 1988 *Electron. Lett.* **24** 308-9.
294. Anderson, B.L., Hall, R.C., Higgins, B.G., Lindsay, G., Stroeve, P. and Kowel, S.T. 1989, *Synth. Met.* **28** D 683-8
295. Stroeve, P., Saperstein, D.D. and Rabolt, J.F. 1990 *J. Chem. Phys.* **92** 6958-67.
296. Young, M.C.J., Lu, W.X., Tredgold, R.H., Hodge, P. and Abbasi, F. 1990 *Electron. Lett.* **26** 993-4.
297. Cresswell, J.P., Tsibouklis, J., Petty, M.C., Feast, W.J., Carr, N., Goodwin, M. and Lvov, Y.M. 1990 *Proc. S.P.I.E. - Int. Soc. Opt. Eng.* **1337** 352-63.
298. Era, M., Nakamura, K., Tsutsui, T., Saito, S., Nino, H., Takehara, K., Isomura, K. and Taniguchi, H. 1990 *Jpn. J. Appl. Phys.* Part 2 **29** L2261-3.
299. Ashwell, G.J., Dawnay, E.J.C., Kuzynski, A.P., and Martin, P.J. 1991 *Proc. SPIE - Int. Soc. Opt. Eng.* **1361** 589-98.
300. Blinov, L.M., Davydova, N.N. Lazarev, V.V. and Yudin, S.G. 1982 *Sov. Tech. Phys. Lett.* **24** 523-5.
301. Smith, G.W., Barton, J.W., Daniel, M.F. and Ratcliffe, N. 1985 *Thin Solid Films* **132** 125-34.
302. Tredgold, R.H., Evans, S.D., Hodge, P. and Hoorfar, A. 1988 *Thin Solid Films* **160** 99-105.
303. Okada, S., Nakanishi, H., Matsuda, H., Kato, M., Abe, T. and Ito, H. 1989 *Thin Solid Films* **178** 313-18.
304. Lvov, Y.M., Troitsky, V.I. and Feigin, L.A. 1989 *Mol. Cryst. Liq. Cryst.* **172** 89-97.
305. Decher, G., Tieke, B., Bosshard, C. and Guenter, P. 1988 *J. Chem. Soc: Chem. Commun.* 933-4.
306. Bosshard, C., Decher, G., Tieke, B. and Guenter, P. 1988 *Proc. S.P.I.E.* **1017** Nonlinear Optical Materials 141-7.
307. Bosshard, C., Kuepfer, M., Florsheimer, M., Guenter, P., Pasquier, C., Zahir, S., and Seifert, M. 1991 *Makromol. Chem. Makromol. Symp.* **46** 27-36.
308. Pasquier, C., Tieke, B., Zahir, S., Bosshard, C. and Guenter, P. 1991 *Chem. Mater.* **3** 211-13.
309. Ashwell, G.J., Dawnay, E.J.C., Kuczynski, A.P., Szablewski, M., Sandy, I.M., Bryce, M.R. , Grainger, A.M. and Hasan, M. 1990 *J. Chem. Soc. Faraday Trans.* **86** 1117-21.
310. Allen, S., Mclean, T.D., Gordon, P.F., Bothwell, B.D., Robin, P. and Ledoux, I. 1988 *Proc. S.P.I.E* **971** 206-15.
311. Ledoux, I., Josse, D., Zyss, J., McLean, T., Gordon, P.F., Hann, R.A. and Allen, S. 1988 *J. Chimie Physique* **85** 1085-90.
312. Selfridge, R.H., Moon, T.K., Stroeve, P., Lam, J.Y.S., Kowel, S.T. and Knoesen, A. 1988 *Proc. S.P.I.E.* **971** 197-205.
313. Tao, F., Xu, L., Chen, G., Yang, X., Wang, G., Zhang, Z. and Wang, W. 1989, *Chin. Sci. Bull.* **34** 1184-9.

314. Ancelin, H., Briody, G., Yarwood, J., Lloyd, J.P., Petty, M.C. Ahmad, M.M. and Feast, W. 1990 *Langmuir* **6** 172–7.
315. Geddes, N.J, Jurich, M.C., Swalen, J.D., Tweig, R. and Rabolt, J.F. 1991 *J. Chem. Phys.* **94** 1603–12.
316. Hsiung, H., Rodriguez-Parada, J. and Bekerbauer, R. 1991 *Chem. Phys. Lett.* **182** 88–92.
317. Penner, T.L., Willand, C.S., Robello, D.R., Schildkraut, J.S. and Ulman, A. 1991, *Proc. S.P.I.E.* **1436** 169–78.
318. Ashwell, G.J., Malhotra, M., Bryce, M.R. and Grainger, A.M. 1991 *Synth. Met.* **43** 3173–6.
319. Hoover, J.M., Henry, R.A., Lindsay, G.A., Lowe-Ma, C.K., Nadler, M.P., Nee, S.M., Seltzer, M.D. and Stenger-Smith, J.D. 1991 *Polymer Preprints* **32** 197–8.
320. Scheelen, A., Winant, P. and Persoons, A. 1991 *NATO ASI Ser.* E.**194** 497–511.
321. Jones, C.A., Petty, M.C. and Roberts, G.G. 1986 *Proc. I.E.E.E. 6th Int. Symp. Appl. Ferroelectricity* 195–8.
322. Smith, G.W. and Evans, T.J. 1987 *Thin Solid Films* **146** 7–13.
323. Novak, V.R., Lvov, Y.M., Myagkov, I.V. and Teternik, G.A. 1987 *J.E.T.P. Lett.* **45** 698–701.
324. Jones, C.A., Petty, M.C., Roberts, G.G., Davies, G, Yarwood, J., Ratcliffe, N.M. and Barton, J.W. 1987 *Thin Solid Films* **155** 187–95.
325. Jones, C.A., Petty, M,C., Davies, G. and Yarwood, J. *J. Phys. D: Appl. Phys.* **21** 95–100.
326. Jones, C.A., Petty, M.C., Russell, G.J. and Roberts, G.G. 1988 *J. Chim. Phys., Phys-Chim. Biol.* **85** 1099–102.
327. Colbrook, R. and Roberts, G.G. 1989 *Thin Solid Films* **179** 335–41.
328. Kamata, T., Umemura, J., Takenaka, T. and Koizumi, N. 1991 *J. Phys. Chem.* **95** 4092–8.
329. Nogami, Y., Hamanaka, H. and Ishiguro, T. 1991 *J. Phys. Soc. Jpn.* **60** 1860–3.
330. Nakanashi, H., Okada, S., Matsuda, H., Kato, M., Sugi, M., Saito, M. and Iizama, S. 1987 *Jpn. J. Appl. Phys.* **26** 1622–4.
331. Lvov, Y.M., Gurskaya, O.B., Berzina, T.S. and Troitskii, V.I. 1989 *Thin Solid Films* **182** 283–96.
332. Matsumoto, M. Nakamurz, T. Tanaka, M. Segiguchi, T. Komizo, H., Matsuzaki, S.Y. Manda, E., Kawabata, Y. and Saito, M. 1987 *Bull. Chem. Soc. Jpn.* **60** 2737–42.
333. Decher, G., Tieke, B., Bosshard, C. and Guenter, P. 1989 *Ferroelectrics* **91** 193–207.
334. Bosshard, C., Tieke, B., Seifert, M. and Guenter, P. 1989 *Inst. Phys. Conf. Ser.* **103** 181–6.
335. Decher, G., Klinkhammer, F., Peterson, I.R. and Seitz, R. 1989 *Thin Solid Films* **178** 445–51.
336. Bosshard, C., Keupfer, M., Guenter, P., Pasquier, C., Zahir, S. and Seifert, M 1990 *Proc. S.P.I.E.* **1273** 70–6.
337. Bosshard, C., Keupfer, M. Guenter, P., Pasquier, C., Zahir, S. and Seifert, M. 1990 *Appl. Phys. Lett.* **56** 1204–6.
338. Mayimato. Y., Kaifu, K., Koyano, T., Saito, M. and Kato, M. 1991 *Jpn. J. Appl. Phys.* Part 2 **30** L1647–9.
339. Bosshard, C., Florsheimer, M., Keupfer, M. and Guenter, P. 1991 *Optics Commun.* **85** 247–53.

340. Nalwa, H.S., Nakajima, K., Watanabe, T. Nakamura, K. Yamada, A. and Miyata, S. 1991 *Jpn. J. Appl. Phys.* Part 1 **30** 983-9.
341. Feibig, K.A. and Dormann, E. 1991 *Appl. Phys.* A **52** 268-72.
342. Sakaguchi, H., Nakemura, H., Nagamura, T., Ogawa, T. and Matsuo, T. 1989 *Chem. Lett.* 1715-3.
343. Allen, S., Hann, R.A., Gupta, S.K., Gordon, P.F., Bothwell, B.D. Ledoux, I., Vidakovic, P., Zyss, J., Robin, P., Chastaing, E. and Dubois,. J-C. 1986 *Proc. S.P.I.E.* **682** 97-102.
344. Hayden, L.M., Kowel, S.T. and Srinivasen, M.P. 1987 *Optics Commun.* **61** 351-5.
345. Stroeve, P., Srinivasen, M.P., Higgins, B.G. and Kowel, S.T. 1987 *Thin Solid Films* **146** 209-220.
346. Ledoux, I., Josse, D., Vidakovik, P., Zyss, J., Hann, R.A., Gordon, P.F., Bothwell, B.D., Gupta, S.K., Allen, S., Robin, P., Chastaing, E. and Dubois, J.C. 1987 *Europhys. Lett.* **3** 803-9.
347. Popovitz-Biro, R., Hill, K., Landau, E.M., Lahav, M., Leiserowitz, L., Sagiv, J., Hsiung, H., Meredith, G.R. and Vanherzeele, H. 1988 *J. Am. Chem. Soc* **110** 2672-4.
348. Tsibouklis, J. Cresswell, J.P., Kalita, N., Pearson, C. Maddaford, P.J., Ancelin, H., Yarwood, J., Goodwin, M.J. and Carr, N. 1989 *J. Phys. D: Appl. Phys.* **22** 1608-12.
349. Popovitz-Biro, R., Hill, K., Shavit, E., Hung, D.J., Lahav, M., Leiserowitz, L., Sagiv, J., Hsiung, H., Meredith, G.R. and Vanherzeele, H. 1990 *J. Am. Chem. Soc.* **112** 2498-506.
350. Ashwell, G.J. 1990 *Thin Solid Films* **186** 155-65.
351. Yamashita, Y., Tanaka, S., Imaeda, K. and Inokuchi, H., 1991 *Chem. Lett.* 1213-16.
352. Nakamura, T., Masumoto, M., Tachibana, H., Tanaka, M., Kawabata, Y. and Miura, Y.F. 1991 *Synth. Met.* **42** 1487-90.
353. Chapman, J.A. and Tabor, D. 1957 *Proc. Roy. Soc.* A **242** 96-107.
354. Allara, D.L. and Nuzzo, R.G. 1985 *Langmuir* **1** 45-52.
355. Allara, D.L. and Nuzzo, R.G. 1985 *Langmuir* **1** 52-66.
356. Chen, S.H. and Frank, C.W. 1989 *Langmuir* **5** 978-87.
357. Sagiv, J. 1980 *J. Am. Chem. Soc.* **102** 92-8.
358. Maoz, R. and Sagiv, J. 1984 *J. Colloid Interface Sci.* **100** 465-96.
359. Gun, J., Iscovici, R. and Sagiv, J. 1984 *J. Colloid Interface Sci.* **101** 201-13
360. Pomerantz, M., Segmuller, A., Netzer, L. and Sagiv, J. 1985 *Thin Solid Films* **132** 153-62.
361. Cohen, S.R., Naaman, R. and Sagiv, J. 1986 *J. Phys. Chem.* **90** 3054-6.
362. Carson, G.A. and Granick, S. 1990 *J. Mater. Res.* **5** 1745-51.
363. Guzonas, D.A., Hair, M.L. and Tripp, C.P. 1990 *Appl. Spectrosc.* **44** 290-3.
364. Ogawa, K., Mino, N., Nakajima, K., Azuma, Y. and Ohmura, T. 1991 *Langmuir* **7** 1473-7.
365. Tilliman, N., Ulman, A. Schildkraut, J.S. and Penner, T.L. 1988 *J. Am. Chem. Soc.* **110** 6136-44.
366. Tillman, N. Ulman, A. and Elman, J.F. 1990 *Langmuir* **6** 1512-8.
367. Nuzzo, R.G. and Allara, D.L. 1983 *J. Am. Chem. Soc.* **105** 4481-3.
368. Porter, M.D., Bright, T.B., Allara, D.L. and Chidsey, C.E.D., 1987 *J. Am. Chem. Soc.* **109** 3559-68.

369. Pashley, D.W. 1959 *Phil. Mag.* **4** 316–24.
370. Pashley, D.W. 1960 *Proc. Roy. Soc.* A **255** 218.
371. Strong, L. and Whitesides, G.M. 1988 *Langmuir* **4** 546–58.
372. Chidsey, C.E.D., Liu, G.Y., Rowntree, P. and Scoles, G. 1989 *J. Chem. Phys.* **91** 4421–3.
373. Ulman, A. Ellers, J. E. and Tillman, N. 1989 *Langmuir* **5** 1147–52.
374. Nuzzo, R.G., Korenic, E.M. and Dubois, L.H. 1990 *J. Chem. Phys.* **93** 767–73.
375. Dubois, L.H., Zegarski, B.R. and Nuzzo, R.G. 1990 *J. Electron. Spectrosc. Related. Phenom.* **54–55** 1143–52.
376. Laibinis, P.E., Whitesides, G.M., Allara, D.L., Tao, Y.T., Parikh, A.N. and Nuzzo, R.G. 1991 *J. Am. Chem. Soc.* **113** 7152–67.
377. Walczak, M.M., Chung, C., Stole, S.M. Widrig, C.A. and Porter, M.D. 1991 *J. Am. Chem. Soc.* **113** 2370–78.
378. Fenter, P., Eisenberger, P., Li, J., Camillone, N., Bernasek, S., Scoles, G., Ramanarayanan, T.A and Liang, K.S. 1991 *Langmuir* **7** 2013–16.
379. Bryant, M.A. and Pemberton, J.E. 1991 *J. Am. Chem. Soc.* **113** 3629–37.
380. Bryant, M.A. and Pemberton, J.E. 1991 *J. Am. Chem. Soc* **113** 8284–93.
381. Troughton, E.B., Bain, C.D., Whitesides, G.M., Nuzzo, R.G., Allara, D.L. and Porter, M.D. 1988 *Langmuir* **4** 365–85.
382. Bain, C.D., Troughton, E.B., Tao, Y.T., Evall, J., Whitesides, G.M. and Nuzzo, R.G. 1989 *J. Am. Chem. Soc.* **111** 321–35.
383. Dubois, L.H., Zegarski, B.R. and Nuzzo, R.G. 1990 *J. Am. Chem. Soc.* **112** 570–9.
384. Nuzzo, R.G., Dubois, L.H. and Allara, D.L. 1990 *J. Am. Chem. Soc.* **112** 558–69.
385. Chidsey, C.E.D. and Loiacono, D.N. 1990 *Langmuir* **6** 682–91.
386. Whitesides, G.E. and Laibinis, P.E. 1990 *Langmuir* **6** 87–96.
387. Evans, S.D., Sharma, R. and Ulman, A. 1991 *Langmuir* **7** 156–61.
388. Tidswell, I, M., Rabedeau, T.A., Pershan, P.S., Folkers, J.P. Baker, M.V. and Whitesides, G.M. 1991 *Phys. Rev.* B **44** 10869–79.
389. Ulman, A., Evans, S.D., Shnidman, Y., Sharma, R., Eilers, J.E. and Chang, J.C. 1991 *J. Am. Chem. Soc.* **113** 1499–1506.
390. Evans, S.D., Urankar, E., Ulman, A. and Ferris, N. 1991 *J. Am. Chem. Soc* **113** 4121–31.
391. Evans, S.D., Goppert-Berarducci, K.E., Urankar, E., Gerenser, L.J. and Ulman, A. 1991 *Langmuir* **7** 2700–9.
392. Lee, H., Kepley, L.J., Hong, H.G. and Mallouk, T.E. 1988 *J. Am. Chem. Soc.* **110** 618–20.
393. Lee, H., Kepley, L.J., Hong, H.G., Akhter, S. and Mallouk, T.E. 1988 *J. Phys. Chem.* **92** 2597–601.
394. Tillman, N., Ulman, A. and Penner, T.L. 1989 *Langmuir* **5** 101–111.
395. Putvinski, T.M., Schilling, M.L., Katz. H.E., Chidsey, C.E.D., Mujsce, A.M. and Emerson, A.B. 1990 *Langmuir* **6** 1567–71.
396. Jin, J.Y. and Johnstone, A.W. 1991 *J. Mater. Chem.* **1** 457–60.
397. Tredgold, R.H., Winter, C.S. and El-Badawy, Z.I. 1985 *Electron. Lett.* **21** 54–55.
398. Banks, R.E. 1964 *Fluorocarbons and Their Derivatives* (Oldbourne Press, London) 14.

399. Kubono, A., Okui, N., Katsufumi, T., Umemoto, S. and Sakai, T. 1991 *Thin Solid Films* **199** 385–93.
400. Ogawa, K., Mino, N., Tamura, H. and Hatada, M. 1989 *Jpn. J. Appl. Phys.* **28** L1854–6.
401. Ogawa, K., Mino, N., Tamura, H. and Hatada, M. 1990 *Langmuir* **6** 851–6.
402. Gray, G.W. and Goodby, J.W.G. 1984 *Smectic Liquid Crystals* (Leonard Hill, Glasgow and London).
403. Leadbetter, A.J. 1987 in *Thermotropic Liquid Crystals* edited by G.W. Gray (Wiley, Chichester/New York/Brisbane/Toronto/Singapore)
404. Weiss, P. 1907 *J. Physique* **6** 667.
405. Onsager, L. 1944, *Phys. Rev.* **65** 117–49.
406. Zannoni, C. 1979 in *The Molecular Physics of Liquid Crystals* edited by G.R. Luckhurst and G.W. Gray (Academic Press, London and New York).
407. Allen, M.P. and Wilson, M.R. 1989 *J. of Computer-Aided Mol. Design* **3** 335–53.
408. Verlet, L. 1967 *Phys. Rev.* **159** 98.
409. Frenkel, D., Mulder, B.M. and McTague, J.P. 1984 *Phys. Rev. Lett.* **52** 287.
410. Allen, M.P., Frenkel, D. and Talbot, J, 1989 *J. Comput. Phys. Rep.* **9** 301.
411. Frenkel, D., Lekkerkerker, H.N.W. and Stroobants, A. 1988 *Nature* **332** 822.
412. Frenkel, D., 1989 *Liq. Cryst.* **5** 929.
413. Luckhurst, G.R., Phippen, R.W. and Stephens, R.A. 1990 *Liq. Cryst.* **8** 451–64.
414. Bareman, J.P. and Klein, M.L 1990 *J. Phys. Chem.* **94** 5202–5.
415. Moller, M.A., Tildesley, D.J., Kim, K.S. and Quirke, N. 1991 *J. Chem. Phys.* **94** 8390–401.
416. Coates, D. 1987 in *Thermotropic Liquid Crystals*, edited by G.W. Gray (Wiley, Chichester/New York/Brisbane/Toronto/Singapore).
417. Tredgold, R.H. and Ali-Adib, Z. 1988 *J. Phys. D: Appl. Phys.* **21** 1467–8.
418. Ali-Adib, Z., Davidson, K., Nooshin, H. and Tredgold, R.H. 1991 *Thin Solid Films* **201** 187–95.
419. Amador, S., Pershan, P.S, Stragier, H., Swanson, B.D., Tweet, D.J., Sorensen, L.B., Sirota, E.B., Ice, G.E. and Habenschuss, A. 1989 *Phys. Rev.* A **39** 2703–8.
420. Decher, G., Maclennan, J., Sohling, U. and Reibel, J. 1992 *Thin Solid Films* **210/211** 504–7.
421. Agarwal, V.K., Igasaki, Y. and Mitsuhashi, H. 1976 *Thin Solid Films* **33** L31.
422. Agarwal, V.K., Igasaki, Y. and Mitsuhashi, H. 1976 *Jpn. J. Appl. Phys.* **15** 2327–32.
423. Barraud, A., Rosilio, C. and Ruaudel-Teixier, A. 1980 *Thin Solid Films* **68** 7–12.
424. Hoshi, H., Dann, A.J. and Maruyama, Y. 1990 *J. Appl. Phys.* **67** 6871–5.
425. Gorter, E. and Grendel, F. 1925 *J. Exp. Med.* **41** 439.
426. Born, M. 1920 *Z. Physik.* **1** 45.

427. Agre, P. and Parker, J.C. 1989 *Red Blood Cell Membranes* (Marcel Dekker, New York).
428. Crick, F. 1988 *What Mad Pursuit* (Penguin Books, London).
429. Stoeckenius, W. and Rowen, R. 1967 *J. Cell Biol.* **34** 65–93.
430. Stoeckenius, W. and Kunau, W.H. 1968 *J. Cell Biol.* **38** 337–57.
431. Oesterhelt, D. and Stoeckenius, W. 1971 *Nature New Biol.* **223** 149–52.
432. Blaurock, A.E. and Stoeckenius, W. 1971 *Nature New Biol.* **223** 152–5.
433. Racker, E. and Stoeckenius, W. 1974 *J. Biol. Chem.* **249** 662–3.
434. Oesterhelt, D. and Stoeckenius, W. 1973 *Proc. Nat. Acad. Sci. U.S.A.* **70** 2853–7.
435. Mitchell, P. 1972 *J. Bioenergetics* **3** 5–24.
436. Unwin, P.T.N. and Henderson, R. 1975 *J. Mol. Biol.* **94** 425–40.
437. Henderson, R. and Unwin, P.T.N. 1975 *Nature* **257** 28–32.
438. Blaurock, A.E. and King, G.I. 1977 *Science* **196** 1101–4.
439. Hayward, S.B., Grano, D.A., Glaesner, R.M. and Fisher, K.A. 1978 *Proc. Nat Acad. Sci. U.S.A.* **75** 4320–24.
440. Silver, B.L. 1985 *The Physical Chemistry of Membranes* (Allen and Unwin, Boston).
441. Seelig, J. 1977 *Q. Rev. Biophys.* **10** 353.
442. Helm, C.A. Möhwald, H., Kjaer, K, and Als-Nielsen, J. 1987 *Biophys. J.* **52** 381–90.
443. Helm, C.A. Möhwald, H. Kjaer, K. and Als-Nielsen, J. 1987 *Europhys. Lett.* **4** 697–703.
444. Möhwald, H., Kenn, R.M., Degenhardt, D., Kjaer, K. and Als-Nielsen, J. 1990 *Physica* A **168** 127–39,
445. Hui, S.W., Parsons, D.F. and Cowden, M. 1974 *Proc. Nat. Acad. Sci. U.S.A.* **71** 5068–72.
446. Fischer, A. and Sackmann, E. 1984 *J. Physique.* **45** 517–27.
447. Bayeri, T.M. Thomas, R.K. Penfold, J. Rennie, A. and Sackmann, E. 1990 *Biophys. J.* **57** 1095–8.
448. Vaknin, D., Kjaer, K., Als-Nielsen, J. and Lösche, M. 1991 *Biophys. J.* **59** 1325–32.
449. Hunt, R.D., Mitchell, M.L. and Dluhy, R.H. 1989 *J. Mol. Struct.* **214** 93–109.
450. Mitchell, M.L. and Dluhy, R.A. 1989 *Proc. S.P.I.E. - Int. Soc. Opt. Eng.* **1145** 255–6.
451. Lösche, M., Duwe, H. P. and Möhwald, H. 1988 *J. Colloid Interface Sci.* **126** 432–44.
452. Florsheimer, M. and Möhwald, H. 1989 *Chem. Phys. Lipids* **49** 231–41.
453. Deitrich, A., Möhwald, H., Rettig, W. and Brezesinski, G. 1991 *Langmuir* **7** 539–46.
454. Ahlers, M., Blankenburg, R., Grainger, D.W., Meller, P., Ringsdorf, H. and Salesse, C. 1989 *Thin Solid Films* **180** 93–9.
455. Ahlers, M., Muller, W., Reichert, A., Ringsdorf, H. and Venzmer, J. 1990 *Angew. Chem. Int. Ed. Eng.* **29** 1269–85.
456. Herron, J.N., Muller, W., Paudler, M., Riegler, H., Ringsdorf, H. and Suci, P.A. 1992 *Langmuir* **8** 1413–16.
457. Ebersole, R.C., Miller, J.A., Moran, J.R. and Ward, M.D. 1990 *J. Am. Chem. Soc.* **112** 3239–41.

458. Haeussling, L., Michel, Bruno., Ringsdorf, H. and Rohrer, H. 1991 *Angew. Chem. Int. Ed. Eng.* **30** 569–72.
459. Haeussling, L., Knoll, W., Ringsdorf, H., Schmitt, F.J. and Yang, J. 1991 *Makromol. Chem. Makromol. Symp.* **46** 145–55.
460. Morgan, H., Taylor, D.M. and D'Silva, C. 1992 *Thin Solid Films* **209** 122–5.
461. Taylor, D.M., Morgan, H, and D'Silva, C. 1991 *J. Phys. D: Appl. Phys.* **24** 1443–50.
462. Morgan, H., Taylor, D.M., D'Silva, C. and Fukushima, H. 1992 *Thin Solid Films* **210/211** 773–5.
463. Garnaes, J., Schwartz, D.K., Viswanathan, R. and Zasadzinski, A.N. 1992 *Nature* **357** 54–7.
464. Chi, L.F., Eng, L.M., Graf, K. and Fuchs, H. 1992 *Langmuir* **8** 2255–61.
465. Ali-Adib, Z., Hodge, P., Tredgold, R.H., Woolley, M. and Pidduck, A.J. 1993 to be published.

Index